The Molecule

"The Building Blocks of Life"

Edited by Paul F. Kisak

Contents

Chapter 1

Molecule

For the scientific journal, see Molecules (journal).

A **molecule** (/ˈmɒlɪkjuːl/ from Latin moles "mass"[1]) is an electrically neutral group of two or more atoms held together

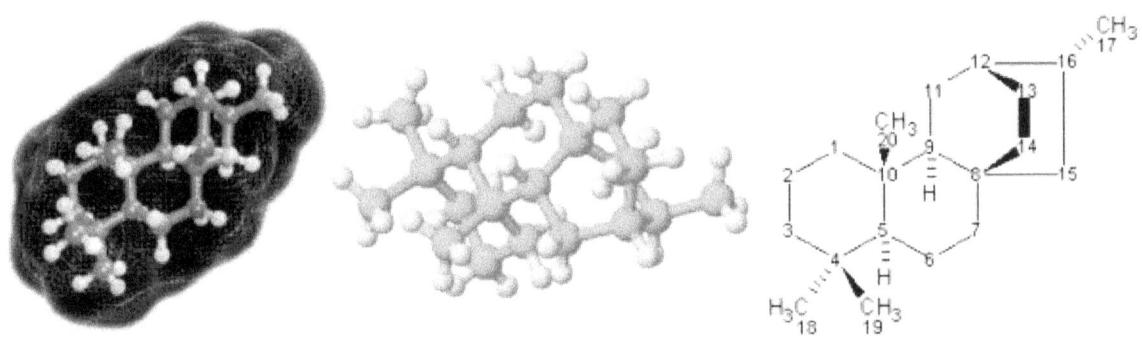

3D (left and center) and 2D (right) representations of the terpenoid molecule atisane

by chemical bonds.[2][3][4][5][6] Molecules are distinguished from ions by their lack of electrical charge. However, in quantum physics, organic chemistry, and biochemistry, the term *molecule* is often used less strictly, also being applied to polyatomic ions.

In the kinetic theory of gases, the term *molecule* is often used for any gaseous particle regardless of its composition. According to this definition, noble gas atoms are considered molecules despite being composed of a single non-bonded atom.[7]

A molecule may be homonuclear, that is, it consists of atoms of a single chemical element, as with oxygen (O_2); or it may be heteronuclear, a chemical compound composed of more than one element, as with water (H_2O). Atoms and complexes connected by non-covalent bonds such as hydrogen bonds or ionic bonds are generally not considered single molecules.[8]

Molecules as components of matter are common in organic substances (and therefore biochemistry). They also make up most of the oceans and atmosphere. However, the majority of familiar solid substances on Earth, including most of the minerals that make up the crust, mantle, and core of the Earth, contain many chemical bonds, but are *not* made of identifiable molecules. Also, no typical molecule can be defined for ionic crystals (salts) and covalent crystals (network solids), although these are often composed of repeating unit cells that extend either in a plane (such as in graphene) or three-dimensionally (such as in diamond, quartz, or sodium chloride). The theme of repeated unit-cellular-structure also holds for most condensed phases with metallic bonding, which means that solid metals are also not made of molecules. In glasses (solids that exist in a vitreous disordered state), atoms may also be held together by chemical bonds without presence of any definable molecule, but also without any of the regularity of repeating units that characterizes crystals.

1.1   Molecular science

The science of molecules is called *molecular chemistry* or *molecular physics*, depending on whether the focus is on chemistry or physics. Molecular chemistry deals with the laws governing the interaction between molecules that results in the formation and breakage of chemical bonds, while molecular physics deals with the laws governing their structure and properties. In practice, however, this distinction is vague. In molecular sciences, a molecule consists of a stable system (bound state) composed of two or more atoms. Polyatomic ions may sometimes be usefully thought of as electrically charged molecules. The term *unstable molecule* is used for very reactive species, i.e., short-lived assemblies (resonances) of electrons and nuclei, such as radicals, molecular ions, Rydberg molecules, transition states, van der Waals complexes, or systems of colliding atoms as in Bose–Einstein condensate.

1.2   History and etymology

Main article: History of molecular theory

According to Merriam-Webster and the Online Etymology Dictionary, the word "molecule" derives from the Latin "moles" or small unit of mass.

- **Molecule** (1794) – "extremely minute particle", from Fr. *molécule* (1678), from modern Latin. *molecula*, diminutive of Latin *moles* "mass, barrier". A vague meaning at first; the vogue for the word (used until the late 18th century only in Latin form) can be traced to the philosophy of Descartes.

The definition of the molecule has evolved as knowledge of the structure of molecules has increased. Earlier definitions were less precise, defining molecules as the smallest particles of pure chemical substances that still retain their composition and chemical properties.[9] This definition often breaks down since many substances in ordinary experience, such as rocks, salts, and metals, are composed of large crystalline networks of chemically bonded atoms or ions, but are not made of discrete molecules.

1.3   Molecular size

Most molecules are far too small to be seen with the naked eye, but there are exceptions. DNA, a macromolecule, can reach macroscopic sizes, as can molecules of many polymers. Molecules commonly used as building blocks for organic synthesis have a dimension of a few angstroms (Å) to several dozen Å. Single molecules cannot usually be observed by light (as noted above), but small molecules and even the outlines of individual atoms may be traced in some circumstances by use of an atomic force microscope. Some of the largest molecules are macromolecules or supermolecules.

1.3.1   Smallest molecule diameter

The smallest molecule is the diatomic hydrogen (H_2), with a bond length of 0.74 Å.[10]

1.3.2   Largest molecule diameter

Mesoporous silica have been produced with a diameter of 1000 Å (100 nm)[11]

1.3.3   Radius

Effective molecular radius is the size a molecule displays in solution.[12][13] The table of permselectivity for different substances contains examples.

John Dalton

1.4 Formulas for molecules

1.4.1 Chemical formula types

Main article: Chemical formula

The chemical formula for a molecule uses a single line of chemical element symbols, numbers, and sometimes also other symbols, such as parentheses, dashes, brackets, and *plus* (+) and *minus* (−) signs. These are limited to a single typographic line of symbols, which may include subscripts and superscripts.

A compound's empirical formula is a very simple type of chemical formula. It is the simplest integer ratio of the chemical elements that constitute it. For example, water is always composed of a 2:1 ratio of hydrogen to oxygen atoms, and ethyl alcohol or ethanol is always composed of carbon, hydrogen, and oxygen in a 2:6:1 ratio. However, this does not determine the kind of molecule uniquely – dimethyl ether has the same ratios as ethanol, for instance. Molecules with the same atoms in different arrangements are called isomers. Also carbohydrates, for example, have the same ratio (carbon:hydrogen:oxygen = 1:2:1) (and thus the same empirical formula) but different total numbers of atoms in the molecule.

The molecular formula reflects the exact number of atoms that compose the molecule and so characterizes different molecules. However different isomers can have the same atomic composition while being different molecules.

The empirical formula is often the same as the molecular formula but not always. For example, the molecule acetylene has molecular formula C_2H_2, but the simplest integer ratio of elements is CH.

The molecular mass can be calculated from the chemical formula and is expressed in conventional atomic mass units equal to 1/12 of the mass of a neutral carbon-12 (^{12}C isotope) atom. For network solids, the term formula unit is used in stoichiometric calculations.

1.4.2 Structural formula

Main article: Structural formula

For molecules with a complicated 3-dimensional structure, especially involving atoms bonded to four different substituents, a simple molecular formula or even semi-structural chemical formula may not be enough to completely specify the molecule. In this case, a graphical type of formula called a structural formula may be needed. Structural formulas may in turn be represented with a one-dimensional chemical name, but such chemical nomenclature requires many words and terms which are not part of chemical formulas.

1.5 Molecular geometry

Main article: Molecular geometry

Molecules have fixed equilibrium geometries—bond lengths and angles— about which they continuously oscillate through vibrational and rotational motions. A pure substance is composed of molecules with the same average geometrical structure. The chemical formula and the structure of a molecule are the two important factors that determine its properties, particularly its reactivity. Isomers share a chemical formula but normally have very different properties because of their different structures. Stereoisomers, a particular type of isomers, may have very similar physico-chemical properties and at the same time different biochemical activities.

1.6 Molecular spectroscopy

Main article: Spectroscopy

Molecular spectroscopy deals with the response (spectrum) of molecules interacting with probing signals of known energy (or frequency, according to Planck's formula). Molecules have quantized energy levels that can be analyzed by detecting the molecule's energy exchange through absorbance or emission.[14] Spectroscopy does not generally refer to diffraction studies where particles such as neutrons, electrons, or high energy X-rays interact with a regular arrangement of molecules (as in a crystal).

1.7 Theoretical aspects

The study of molecules by molecular physics and theoretical chemistry is largely based on quantum mechanics and is essential for the understanding of the chemical bond. The simplest of molecules is the hydrogen molecule-ion, H_2^+, and the simplest of all the chemical bonds is the one-electron bond. H_2^+ is composed of two positively charged protons and one negatively charged electron, which means that the Schrödinger equation for the system can be solved more easily due to the lack of electron–electron repulsion. With the development of fast digital computers, approximate solutions for more complicated molecules became possible and are one of the main aspects of computational chemistry.

When trying to define rigorously whether an arrangement of atoms is "sufficiently stable" to be considered a molecule, IUPAC suggests that it "must correspond to a depression on the potential energy surface that is deep enough to confine at least one vibrational state".[2] This definition does not depend on the nature of the interaction between the atoms, but only on the strength of the interaction. In fact, it includes weakly bound species that would not traditionally be considered molecules, such as the helium dimer, He_2, which has one vibrational bound state[15] and is so loosely bound that it is only likely to be observed at very low temperatures.

Whether or not an arrangement of atoms is "sufficiently stable" to be considered a molecule is inherently an operational definition. Philosophically, therefore, a molecule is not a fundamental entity (in contrast, for instance, to an elementary particle); rather, the concept of a molecule is the chemist's way of making a useful statement about the strengths of atomic-scale interactions in the world that we observe.

1.8 See also

- Atom

- Van der Waals molecule

- Diatomic molecule

- Small molecule

- Chemical polarity

- Molecular geometry

- Covalent bond

- Noncovalent bonding

- list of compounds for a list of chemical compounds

- List of molecules in interstellar space

- Software for molecular mechanics modeling

- Molecular Hamiltonian

- Molecular ion

- Molecular orbital

- Molecular modelling

- Molecular design software

- WorldWide Molecular Matrix

- Periodic Systems of Small Molecules

1.9   References

[1] http://www.etymonline.com/index.php?term=molecule

[2] Template:GoldBookRefwhen they pook

[3] Ebbin, Darrell D. (1990). *General Chemistry* (3rd ed.). Boston: Houghton Mifflin Co. ISBN 0-395-43302-9.

[4] Brown, T.L.; Kenneth C. Kemp; Theodore L. Brown; Harold Eugene LeMay et al. (2003). *Chemistry – the Central Science* (9th ed.). New Jersey: Prentice Hall. ISBN 0-13-066997-0.

[5] Chang, Raymond (1998). *Chemistry* (6th ed.). New York: McGraw Hill. ISBN 0-07-115221-0.

[6] Zumdahl, Steven S. (1997). *Chemistry* (4th ed.). Boston: Houghton Mifflin. ISBN 0-669-41794-7.

[7] Chandra, Sulekh (2005). *Comprehensive Inorganic Chemistry*. New Age Publishers. ISBN 81-224-1512-1.

[8] Molecule, *Encyclopædia Britannica* on-line

[9] Molecule Definition (Frostburg State University)

[10] Roger L. DeKock; Harry B. Gray; Harry B. Gray (1989). *Chemical structure and bonding*. University Science Books. p. 199. ISBN 0-935702-61-X.

[11] http://pubs.acs.org/doi/abs/10.1021/ac303274w

[12] Chang RL; Deen WM; Robertson CR; Brenner BM. (1975). "Permselectivity of the glomerular capillary wall: III. Restricted transport of polyanions". *Kidney Int.* **8** (4): 212–218. doi:10.1038/ki.1975.104. PMID 1202253.

[13] Chang RL; Ueki IF; Troy JL; Deen WM et al. (1975). "Permselectivity of the glomerular capillary wall to macromolecules. II. Experimental studies in rats using neutral dextran". *Biophys J.* **15** (9): 887–906. Bibcode:1975BpJ....15..887C.doi:10.1016/S0 3495(75)85863-2. PMC 1334749. PMID 1182263.

[14] IUPAC, *Compendium of Chemical Terminology*, 2nd ed. (the "Gold Book") (1997). Online corrected version: (1997,2006) "spectroscopy".

[15] Anderson JB (May 2004). "Comment on "An exact quantum Monte Carlo calculation of the helium-helium intermolecular potential" [J. Chem. Phys. 115, 4546 (2001)]". *J Chem Phys* **120** (20): 9886–7. Bibcode:2004JChPh.120.9886A. doi:10.1063/1.1704638. PMID 15268005.

1.10   External links

- Molecule of the Month – School of Chemistry, University of Bristol

Chapter 2

Molecular mass

Molecular mass or **molecular weight** is the mass of a molecule. It is calculated as the sum of the mass of each constituent atom multiplied by the number of atoms of that element in the molecular formula. The molecular mass of small to medium size molecules, measured by mass spectrometry, determines stoichiometry. For large molecules such as proteins, methods based on viscosity and light-scattering can be used to determine molecular mass when crystallographic data are not available.

2.1 Definitions

Both atomic and molecular masses are usually obtained relative to the mass of the isotope ^{12}C (carbon 12), which by definition[1] is equal to 12. For example, the molecular weight of methane, whose molecular formula is CH_4, is calculated as follows.

A more proper term would be "relative molecular mass". However the adjective 'relative' is omitted as it is universally assumed that atomic and molecular masses are relative to the mass of ^{12}C. Relative atomic and molecular mass values are dimensionless but are given the "unit" Dalton (formerly atomic mass unit) to indicate that the number is equal to the mass of one molecule divided by $^1/_{12}$ of the mass of one atom of ^{12}C. The mass of 1 mol of substance is designated as molar mass. By definition, it has the unit gram.

In the example above the atomic weight of carbon is given as 12.011, not 12. This is because naturally occurring carbon is a mixture of the isotopes ^{12}C, ^{13}C and ^{14}C which have relative atomic masses of 12, 13 and 14 respectively. Moreover, the proportion of the isotopes varies between samples, so 12.011 is an average value. By contrast, there is less variation in naturally occurring hydrogen so the average atomic weight is known more precisely. The precision of the molecular mass is determined by precision of the least precise atomic mass value, in this case that of carbon. In high-resolution mass spectrometry the isotopomers $^{12}C^1H_4$ and $^{13}C^1H_4$ are observed as distinct molecules, with molecular weights of 16 and 17, respectively. The intensity of the mass-spectrometry peaks is proportional to the isotopic abundances in the molecular species. $^{12}C\,^2H\,^1H_3$ can also be observed with molecular weight of 17.

2.2 Determination of molecular mass

2.2.1 Mass spectrometry

Main article: Mass spectrometry

In mass spectrometry, the molecular mass of a small molecule is usually reported as the monoisotopic mass, that is, the mass of the molecule containing only the most common isotope of each element. Note that this also differs subtly from the molecular mass in that the choice of isotopes is defined and thus is a single specific molecular mass of the many possible. The masses used to compute the monoisotopic molecular mass are found on a table of isotopic masses and are not found on a typical periodic table. The average molecular mass is often used for larger molecules since molecules with many atoms are unlikely to be composed exclusively of the most abundant isotope of each element. A theoretical average molecular mass can be calculated using the relative atomic masses found on a typical periodic table, since there is likely to be a statistical distribution of atoms representing the isotopes throughout the molecule. This however may differ from the true average molecular mass of the sample due to natural (or artificial) variations in the isotopic distributions.

2.2.2   Hydrodynamic methods

To a first approximation, the basis for determination of molecular weight according to Mark–Houwink relations[2] is the fact that the intrinsic viscosity of solutions (or suspensions) of macromolecules depends on volumetric proportion of the dispersed particles in a particular solvent. Specifically, the hydrodynamic size as related to molecular weight depends on a conversion factor, describing the shape of a particular molecule. This allows the apparent molecular weight to be described from a range of techniques sensitive to hydrodynamic effects, including DLS, SEC (also known as GPC) and viscometry. The apparent hydrodynamic size can then be used to approximate molecular weight using a series of macromolecule-specific standards. As this requires calibration, it's frequently described as a "relative" molecular weight determination method.

2.2.3   Static light scattering

It is also possible to determine absolute molecular weight directly from light scattering, traditionally using the Zimm method. This can be accomplished either via classical static light scattering or via multiangle light scattering detectors. Molecular weights determined by this method do not require calibration, hence the term "absolute". The only external measurement required is refractive index increment, which describes the change in refractive index with concentration.

2.3   See also

- Absolute molar mass

- Molar mass distribution

- Dumas method of molecular weight determination

2.4   References

[1] International Union of Pure and Applied Chemistry (1980). "Atomic Weights of the Elements 1979" (PDF). *Pure Appl. Chem.* **52** (10): 2349–84. doi:10.1351/pac198052102349.

[2] Paul, Hiemenz C., and Lodge P. Timothy. Polymer Chemistry. Second ed. Boca Raton: CRC P, 2007. 336, 338–339.

2.5   External links

- A Free Android application for molecular and reciprocal weight calculation of any chemical formula

- Stoichiometry Add-In for Microsoft Excel for calculation of molecular weights, reaction coefficients and stoichiometry.

Chapter 3

Atomic mass unit

The **unified atomic mass unit** (symbol: **u**) or **dalton** (symbol: **Da**) is the standard unit that is used for indicating mass on an atomic or molecular scale (atomic mass). One unified atomic mass unit is approximately the mass of one nucleon (either a single proton or neutron) and is equivalent to 1 g/mol.[1] It is defined as one twelfth of the mass of an unbound neutral atom of carbon-12 in its nuclear and electronic ground state,[2] and has a value of $1.660538921(73) \times 10^{-27}$ kg.[3] The CIPM has categorised it as a non-SI unit accepted for use with the SI, and whose value in SI units must be obtained experimentally.[2]

The **amu** without the "unified" prefix is technically an obsolete unit based on oxygen, which was replaced in 1961. However, some sources may still use the term "amu" but now define it in the same way as u (i.e. based on carbon-12). In this sense, most uses of the terms "atomic mass units" and "amu" today actually refer to unified atomic mass units. For standardization a specific atomic nucleus (carbon-12 vs. oxygen-16) had to be chosen because average mass of a nucleon depends on the count of the nucleons in the atomic nucleus due to mass defect. This is also why the mass of a proton (or neutron) by itself is more than (and not equal to) 1 u.

Atomic mass unit does *not* stand for the unit of mass in the atomic units system, which is rather m_e.

3.1 History

The relative atomic mass (atomic weight) scale has traditionally been a relative scale, that is without an explicit unit, with the first relative atomic mass basis suggested by John Dalton in 1803 as ^{1}H.[4] Despite the initial mass of ^{1}H being used as the natural unit for relative atomic mass, it was suggested by Wilhelm Ostwald that relative atomic mass would be best expressed in terms of units of 1/16 mass of oxygen. This evaluation was made prior to the discovery of the existence of elemental isotopes, which occurred in 1912.[4]

The discovery of isotopic oxygen in 1929 led to a divergence in relative atomic mass representation, with isotopically weighted oxygen (i.e., naturally occurring oxygen relative atomic mass) given a value of exactly 16 atomic mass units (amu) in chemistry, while pure ^{16}O (oxygen-16) was given the mass value of exactly 16 amu in physics.

The divergence of these values could result in errors in computations, and was unwieldy. The chemistry amu, based on the relative atomic mass (atomic weight) of natural oxygen (including the heavy naturally-occurring isotopes ^{17}O and ^{18}O), was about 1.000282 as massive as the physics amu, based on pure isotopic ^{16}O.

For these and other reasons, the reference standard for both physics and chemistry was changed to carbon-12 in 1961.[5] The choice of carbon-12 was made to minimise further divergence with prior literature.[4] The new and current unit was referred to as the "unified atomic mass unit" u.[6] and given a new symbol, "u," which replaced the now deprecated "amu" that had been connected to the old oxygen-based system. The Dalton (Da) is another name for the unified atomic mass unit.[7]

Despite this change, modern sources often still use the old term "amu" but define it as u (1/12 of the mass of a carbon-12 atom), as mentioned in the article's introduction. Therefore, in general, "amu" likely does not refer to the old oxygen standard unit, unless the source material originates from or before the 1960s.

The unified atomic mass unit u was defined as:

$$1\text{u} = m_\text{u} = \frac{1}{12}m\left(^{12}\text{C}\right)$$

3.2 Terminology

The unified atomic mass unit and the dalton are different names for the same unit of measure. As with other unit names such as watt and newton, "dalton" is not capitalized in English, but its symbol Da is capitalized. With the introduction of the name "dalton", there has been a gradual change towards using that name in preference to the name "unified atomic mass unit":

- In 1993, the International Union of Pure and Applied Chemistry approved the use of the dalton with the qualification that the CGPM had not given its approval.[8]

- In 2003 the Consultative Committee for Units, part of the CIPM, recommended a preference for the usage of the "*dalton*" over the "*unified atomic mass unit*" as it "*is shorter and works better with prefixes*".[9]

- In 2005, the International Union of Pure and Applied Physics endorsed the use of the dalton as an alternative to the unified atomic mass unit.[10]

- In 2006, in the 8th edition of the formal definition of SI, the CIPM cataloged the dalton alongside the unified atomic mass unit as a "Non-SI units whose values in SI units must be obtained experimentally: Units accepted for use with the SI".[2] The definition also noted that "*The dalton is often combined with SI prefixes ...*"

- In 2009, when the International Organization for Standardization published updated versions of ISO 80000, it gave mixed messages as to whether or not the unified atomic mass unit had been deprecated: ISO ISO 80000-1:2009 (General), identified the dalton as having "*earlier [been] called the unified atomic mass unit u*",[11] but ISO 80000-10:2009 (atomic and nuclear physics) catalogued both as being alternatives for each other.[12]

- The 2010 version of the Oxford University Press style guide for authors in life sciences gave the following guidance "*Use the Système international d'unités (SI) wherever possible ... The dalton (Da) or more conveniently the kDa is a permitted non-SI unit for molecular mass or mass of a particular band in a separating gel.*"[13] At the same time, the author guidelines for the journal "*Rapid Communications in Mass Spectrometry*" stated "*The dalton (Da) is a unit of mass normally used for the molecular weight ... use of the Da in place of the u has become commonplace in the mass spectrometry literature ... The "atomic mass unit", abbreviated "amu", is an archaic unit*".[14]

- In 2012, in response to the proposed redefinition of the kilogram, it was proposed that the dalton be redefined as being 0.001/NA kg, thereby breaking the link with ^{12}C. This would result in the dalton and the atomic mass unit having slightly different definitions, but the suggestion is that the older unit should be superseded by the "new" dalton.[15]

3.3 Relationship to SI

The definition of the mole, an SI base unit, was accepted by the CGPM in 1971 as:

1. The mole is the amount of substance of a system which contains as many elementary entities as there are atoms in 0.012 kilogram of carbon-12; its symbol is "mol".

2. When the mole is used, the elementary entities must be specified and may be atoms, molecules, ions, electrons, other particles, or specified groups of such particles.

The definition of the mole also determines the value of the universal constant that relates the number of entities to amount of substance for any sample. This constant is called the Avogadro constant, symbol N_A or L, and has the value $6.022140857(74) \times 10^{23}$ mol^{-1} (entities per mole).[16]

Given that the unified atomic mass unit is one twelfth the mass of one atom of carbon-12, meaning the mass of such an atom is 12 u, it follows that there are N_A atoms of carbon-12 in 0.012 kg of carbon-12. This can be expressed mathematically as

N_A (12 u) = 0.012 kg/mol, or

N_A u = 0.001 kg/mol

Masses of proteins are often expressed in daltons. For example, a protein with a molecular weight of 64000 g·mol^{-1} has a mass of 64 kDa.[1]

3.4 Examples

- A hydrogen-1 atom has a mass of 1.0078250 u (1.0078250 Da).

- By definition, a carbon-12 atom has a mass of 12 u (12 Da).

- A molecule of acetylsalicylic acid (Aspirin) has a mass of 180.16 u (180.16 Da).

- Titin, the largest known protein, has an atomic mass of 3-3.7 megadaltons (3000000 Da).[17]

3.5 See also

- Kendrick mass

- Mass-to-charge ratio

- Atomic mass constant

3.6 Notes and references

[1] Stryer, Jeremy M. Berg; John L. Tymoczko; Lubert (2007). "2". *Biochemistry* (6. ed., 3. print. ed.). New York: Freeman. p. 35. ISBN 978-0-7167-8724-2.

[2] International Bureau of Weights and Measures (2006), *The International System of Units (SI)* (PDF) (8th ed.), p. 126, ISBN 92-822-2213-6

[3] Unified Atomic mass unit. Fundamental Physical Constants from NIST

[4] Petley, B. W., "The atomic mass unit", *IEEE Trans. Instrum. Meas.* **38** (2): 175–79, doi:10.1109/19.192268

[5] Holden, Norman E. (2004), "Atomic Weights and the International Committee—A Historical Review", *Chem. Int.* **26** (1): 4–7

[6] IUPAC, *Compendium of Chemical Terminology*, 2nd ed. (the "Gold Book") (1997). Online corrected version: (2006–) "unified atomic mass unit".

[7] IUPAC, *Compendium of Chemical Terminology*, 2nd ed. (the "Gold Book") (1997). Online corrected version: (2006–) "dalton".

[8] Mills, Ian; Cvitaš, Tomislav; Homann, Klaus; Kallay, Nikola; Kuchitsu, Kozo (1993n). *Quantities, Units and Symbols in Physical Chemistry International Union of Pure and Applied Chemistry; Physical Chemistry Division* (PDF) (2nd ed.). International Union of Pure and Applied Chemistry and published for them by Blackwell Science Ltd. ISBN 0-632-03583-8.

[9] "Consultative Committee for Units (CCU); Report of the 15th meeting (17–18 April 2003) to the International Committee for Weights and Measures" (PDF). Retrieved 14 Aug 2010.

[10] "IU14. IUPAC Interdivisional Committee on Nomenclature and Symbols (ICTNS)". Retrieved 2010-08-14.

[11] *International Standard ISO 80000-1:2009 – Quantities and Units – Part 1: General*, International Organization for Standardization, 2009

[12] *International Standard ISO 80000-10:2009 – Quantities and units – Part 10: Atomic and nuclear physics*, International Organization for Standardization, 2009

[13] "Instructions to Authors". *AoB Plants.* Oxford journals; Oxford University Press. Retrieved 2010-08-22.

[14] "Author guidelines". *Rapid Communications in Mass Spectrometry* (Wiley-Blackwell). 2010. Retrieved 2011-05-08.

[15] Leonard, B P (2012). "Why the dalton should be redefined exactly in terms of the kilogram". *Metrologia* **49**: 487–491. Bibcode:2012Metro..49..487L. doi:10.1088/0026-1394/49/4/487.

[16] Mohr, Peter J.; Taylor, Barry N.; Newell, David B. (2008). "CODATA Recommended Values of the Fundamental Physical Constants: 2006". *Rev. Mod. Phys.* **80** (2): 633–730. arXiv:0801.0028. Bibcode:2008RvMP...80..633M. doi:10.1103/RevMod. Direct link to value.

[17] Opitz CA, Kulke M, Leake MC, Neagoe C, Hinssen H, Hajjar RJ, Linke WA (October 2003). "Damped elastic recoil of the titin spring in myofibrils of human myocardium". *Proc. Natl. Acad. Sci. U.S.A.* **100** (22): 12688–93. Bibcode:2003PNAS..10012688O. doi:10.1073/pnas.2133733100. PMC 240679. PMID 14563922.

3.7 External links

- atomic mass unit at sizes.com

Chapter 4

History of molecular theory

Space-filling model of the H_2O molecule.

In chemistry, the **history of molecular theory** traces the origins of the concept or idea of the existence of strong chemical bonds between two or more atoms.

The modern concept of molecules can be traced back towards pre-scientific Greek philosophers such as Leucippus who argued that all the universe is composed of atoms and voids. Circa 450 BC Empedocles imagined fundamental elements (fire ($\triangle$), earth ($\triangledown$), air ($\triangle$), and water ($\triangledown$)) and "forces" of attraction and repulsion allowing the elements to interact. Prior to this, Heraclitus had claimed that fire or change was fundamental to our existence, created through the combination of opposite properties.[1] In the Timaeus, Plato, following Pythagoras, considered mathematical entities such as number, point, line and triangle as the fundamental building blocks or elements of this ephemeral world, and considered the four elements of fire, air, water and earth as states of substances through which the true mathematical principles or elements would pass.[2] A fifth element, the incorruptible quintessence aether, was considered to be the fundamental building block of the heavenly bodies. The viewpoint of Leucippus and Empedocles, along with the aether, was accepted by Aristotle and passed to medieval and renaissance Europe. A modern conceptualization of molecules began to develop in the 19th century along with experimental evidence for pure chemical elements and how individual atoms of different chemical substances such as hydrogen and oxygen can combine to form chemically stable molecules such as water molecules.

4.1   17th century

The earliest views on the shapes and connectivity of atoms was that proposed by Leucippus, Democritus, and Epicurus who reasoned that the solidness of the material corresponded to the shape of the atoms involved. Thus, iron atoms are solid and strong with hooks that lock them into a solid; water atoms are smooth and slippery; salt atoms, because of their taste, are sharp and pointed; and air atoms are light and whirling, pervading all other materials.[3] It was Democritus that was the main proponent of this view. Using analogies based on the experiences of the senses, he gave a picture or an image of an atom in which atoms were distinguished from each other by their shape, their size, and the arrangement of their parts. Moreover, connections were explained by material links in which single atoms were supplied with attachments: some with hooks and eyes others with balls and sockets (see diagram).[4]

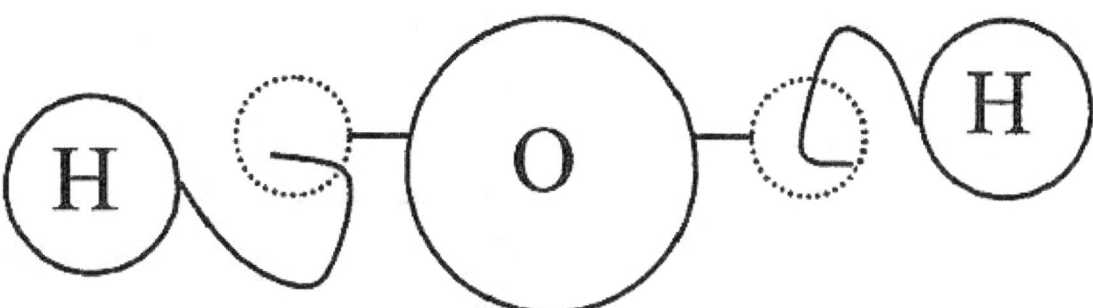

A water molecule as hook-and-eye model might have represented it. Leucippus, Democritus, Epicurus, Lucretius and Gassendi adhered to such conception. Note that the composition of water was not known before Avogadro (c. 1811).

With the rise of scholasticism and the decline of the Roman Empire, the atomic theory was abandoned for many ages in favor of the various four element theories and later alchemical theories. The 17th century, however, saw a resurgence in the atomic theory primarily through the works of Gassendi, and Newton. Among other scientists of that time Gassendi deeply studied ancient history, wrote major works about Epicurus natural philosophy and was a persuasive propagandist of it. He reasoned that to account for the size and shape of atoms moving in a void could account for the properties of matter. Heat was due to small, round atoms; cold, to pyramidal atoms with sharp points, which accounted for the pricking sensation of severe cold; and solids were held together by interlacing hooks.[5] Newton, though he acknowledged the various atom attachment theories in vogue at the time, i.e. "hooked atoms", "glued atoms" (bodies at rest), and the "stick together by conspiring motions" theory, rather believed, as famously stated in "Query 31" of his 1704 *Opticks*, that particles attract one another by some force, which "in immediate contact is extremely strong, at small distances performs the chemical operations, and reaches not far from particles with any sensible effect." [6]

In a more concrete manner, however, the concept of aggregates or units of bonded atoms, i.e. "molecules", traces its origins to Robert Boyle's 1661 hypothesis, in his famous treatise *The Sceptical Chymist*, that matter is composed of *clusters of particles* and that chemical change results from the rearrangement of the clusters. Boyle argued that matter's basic elements consisted of various sorts and sizes of particles, called "corpuscles", which were capable of arranging themselves into groups.

In 1680, using the corpuscular theory as a basis, French chemist Nicolas Lemery stipulated that the acidity of any substance consisted in its pointed particles, while alkalis were endowed with pores of various sizes.[7] A molecule, according to this view, consisted of corpuscles united through a geometric locking of points and pores.

4.2 18th century

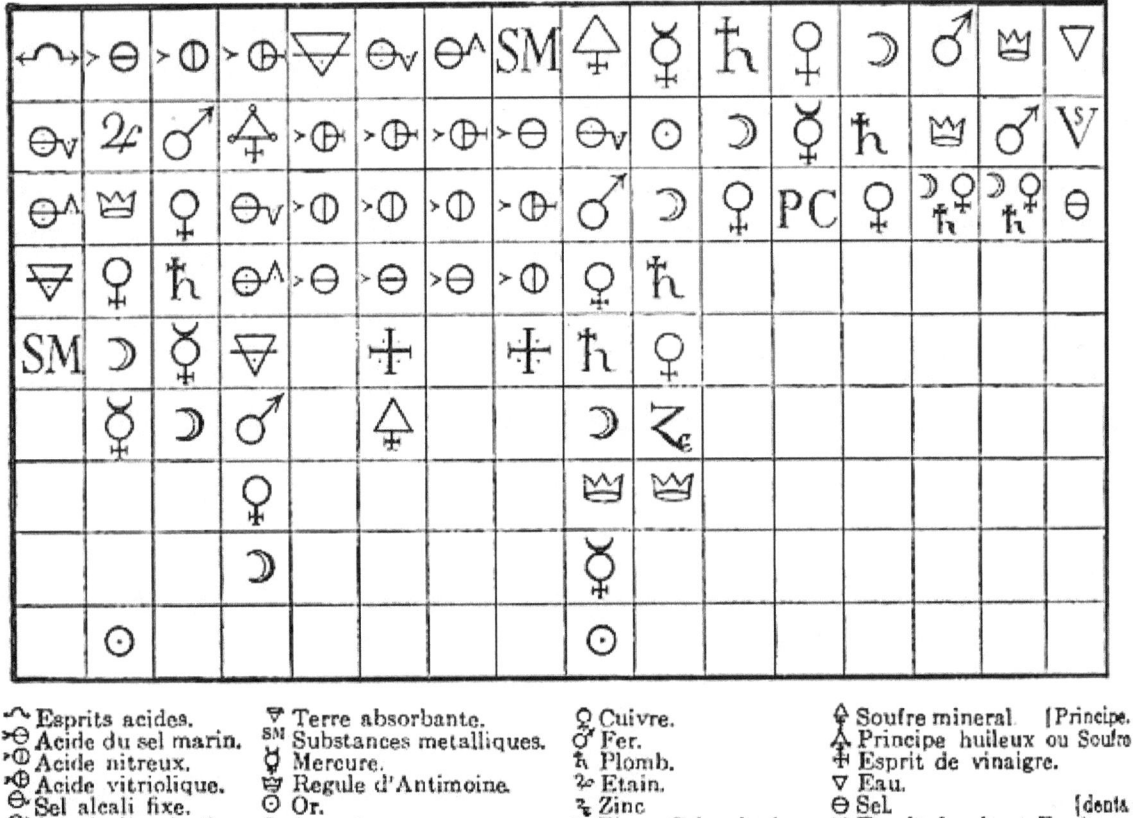

Étienne François Geoffroy's 1718 Affinity Table*: at the head of the column is a substance with which all the substances below can combine.*

An early precursor to the idea of bonded "combinations of atoms", was the theory of "combination via chemical affinity". For example, in 1718, building on Boyle's conception of combinations of clusters, the French chemist Étienne François Geoffroy developed theories of chemical affinity to explain combinations of particles, reasoning that a certain alchemical "force" draws certain alchemical components together. Geoffroy's name is best known in connection with his tables of "affinities" (*tables des rapports*), which he presented to the French Academy in 1718 and 1720.

These were lists, prepared by collating observations on the actions of substances one upon another, showing the varying degrees of affinity exhibited by analogous bodies for different reagents. These tables retained their vogue for the rest of the century, until displaced by the profounder conceptions introduced by CL Berthollet.

In 1738, Swiss physicist and mathematician Daniel Bernoulli published *Hydrodynamica*, which laid the basis for the kinetic theory of gases. In this work, Bernoulli positioned the argument, still used to this day, that gases consist of great numbers of molecules moving in all directions, that their impact on a surface causes the gas pressure that we feel, and that what we experience as heat is simply the kinetic energy of their motion. The theory was not immediately accepted, in part because conservation of energy had not yet been established, and it was not obvious to physicists how the collisions between molecules could be perfectly elastic.

In 1789, William Higgins published views on what he called combinations of "ultimate" particles, which foreshadowed the concept of valency bonds. If, for example, according to Higgins, the force between the ultimate particle of oxygen and the ultimate particle of nitrogen were 6, then the strength of the force would be divided accordingly, and similarly for the other combinations of ultimate particles:

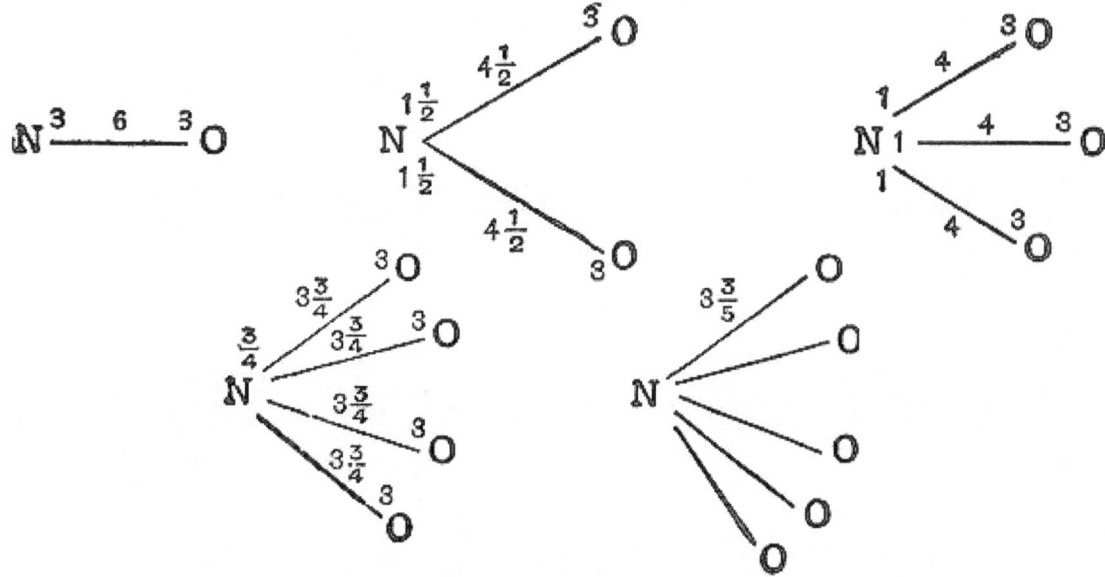

William Higgins' combinations of ultimate particles (1789)

4.3   19th century

Similar to these views, in 1803 John Dalton took the atomic weight of hydrogen, the lightest element, as unity, and determined, for example, that the ratio for nitrous anhydride was 2 to 3 which gives the formula N_2O_3. Interestingly, Dalton incorrectly imagined that atoms "hooked" together to form molecules. Later, in 1808, Dalton published his famous diagram of combined "atoms":

In Amedeo Avogadro's famous 1811 paper "Essay on Determining the Relative Masses of the Elementary Molecules of Bodies", he essentially states, i.e. according to Partington's *A Short History of Chemistry*, that:[8]

> The smallest particles of gases are not necessarily simple atoms, but are made up of a certain number of these atoms united by attraction to form a single **molecule**.

Note that this quote is not a literal translation. Avogadro uses the name "molecule" for both atoms and molecules. Specifically, he uses the name "elementary molecule" when referring to atoms and to complicate the matter also speaks of "compound molecules" and "composite molecules".

During his stay in Vercelli, Avogadro wrote a concise note (*memoria*) in which he declared the hypothesis of what we now call Avogadro's law: *equal volumes of gases, at the same temperature and pressure, contain the same number of molecules*. This law implies that the relationship occurring between the weights of same volumes of different gases, at the same temperature and pressure, corresponds to the relationship between respective molecular weights. Hence, relative molecular masses could now be calculated from the masses of gas samples.

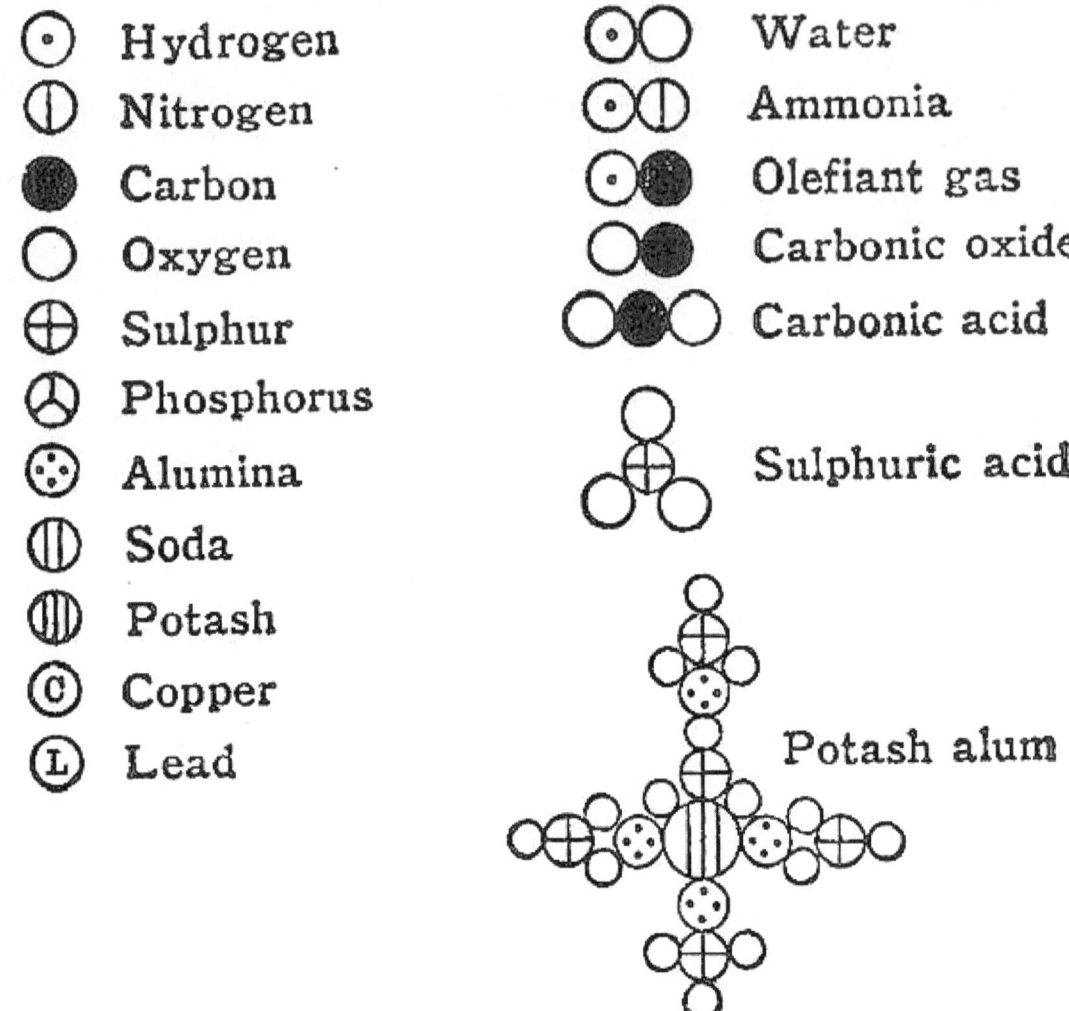

John Dalton's union of atoms combined in ratios (1808)

Avogadro developed this hypothesis in order to reconcile Joseph Louis Gay-Lussac's 1808 law on volumes and combining gases with Dalton's 1803 atomic theory. The greatest difficulty Avogadro had to resolve was the huge confusion at that time regarding atoms and molecules—one of the most important contributions of Avogadro's work was clearly distinguishing one from the other, admitting that simple particles too could be composed of molecules, and that these are composed of atoms. Dalton, by contrast, did not consider this possibility. Curiously, Avogadro considers only molecules containing even numbers of atoms; he does not say why odd numbers are left out.

In 1826, building on the work of Avogadro, the French chemist Jean-Baptiste Dumas states:

> Gases in similar circumstances are composed of **molecules** or atoms placed at the same distance, which is the same as saying that they contain the same number in the same volume.

In coordination with these concepts, in 1833 the French chemist Marc Antoine Auguste Gaudin presented a clear account of Avogadro's hypothesis,[9] regarding atomic weights, by making use of "volume diagrams", which clearly show both semi-correct molecular geometries, such as a linear water molecule, and correct molecular formulas, such as H_2O:

In two papers outlining his "theory of atomicity of the elements" (1857–58), Friedrich August Kekulé was the first to offer a theory of how every atom in an organic molecule was bonded to every other atom. He proposed that carbon atoms

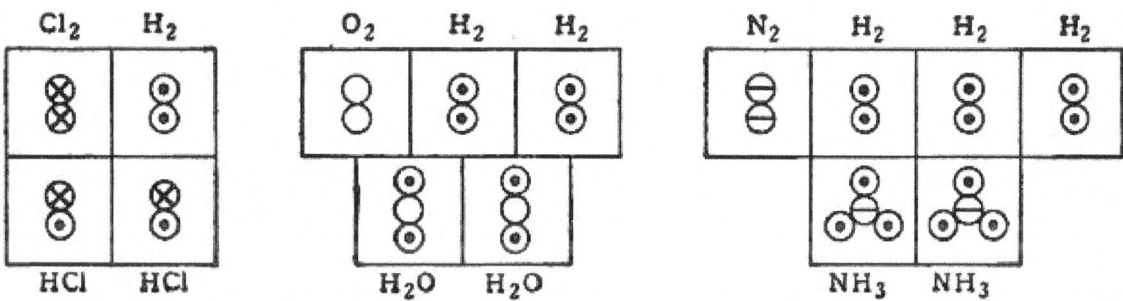

Marc Antoine Auguste Gaudin's volume diagrams of molecules in the gas phase (1833)

were tetravalent, and could bond to themselves to form the carbon skeletons of organic molecules.

In 1856, Scottish chemist Archibald Couper began research on the bromination of benzene at the laboratory of Charles Wurtz in Paris.[10] One month after Kekulé's second paper appeared, Couper's independent and largely identical theory of molecular structure was published. He offered a very concrete idea of molecular structure, proposing that atoms joined to each other like modern-day Tinkertoys in specific three-dimensional structures. Couper was the first to use lines between atoms, in conjunction with the older method of using brackets, to represent bonds, and also postulated straight chains of atoms as the structures of some molecules, ring-shaped molecules of others, such as in tartaric acid and cyanuric acid [11] In later publications, Couper's bonds were represented using straight dotted lines (although it is not known if this is the typesetter's preference) such as with alcohol and oxalic acid below:

$$CH_3$$
$$\vdots$$
$$CH_2 ---OH$$
$$O$$

$$C \begin{matrix} O \\ \overline{O_2} \end{matrix} ---OH$$
$$\vdots$$
$$C \begin{matrix} O \\ \overline{O} \end{matrix} ---OH$$
$$---_2$$

Archibald Couper's molecular structures, for alcohol and oxalic acid, using elemental symbols for atoms and lines for bonds (1858)

In 1861, an unknown Vienna high-school teacher named Joseph Loschmidt published, at his own expense, a booklet entitled *Chemische Studien I*, containing pioneering molecular images which showed both "ringed" structures as well as double-bonded structures, such as:[12]

Loschmidt also suggested a possible formula for benzene, but left the issue open. The first proposal of the modern structure for benzene was due to Kekulé, in 1865. The cyclic nature of benzene was finally confirmed by the crystallographer Kathleen Lonsdale. Benzene presents a special problem in that, to account for all the bonds, there must be alternating double carbon bonds:

In 1865, German chemist August Wilhelm von Hofmann was the first to make stick-and-ball molecular models, which he used in lecture at the Royal Institution of Great Britain, such as methane shown below:

The basis of this model followed the earlier 1855 suggestion by his colleague William Odling that carbon is tetravalent. Hofmann's color scheme, to note, is still used to this day: nitrogen = blue, oxygen = red, chlorine = green, sulfur = yellow,

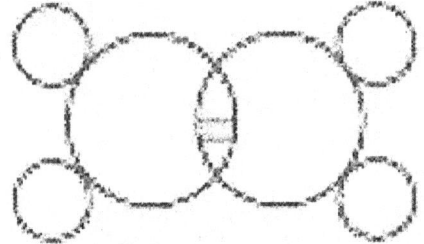 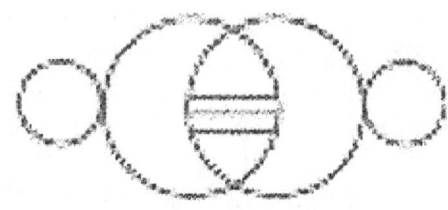

Joseph Loschmidt's molecule drawings of ethylene $H_2C=CH_2$ and acetylene $HC\equiv CH$ (1861)

hydrogen = white.[13] The deficiencies in Hofmann's model were essentially geometric: carbon bonding was shown as planar, rather than tetrahedral, and the atoms were out of proportion, e.g. carbon was smaller in size than the hydrogen.

In 1864, Scottish organic chemist Alexander Crum Brown began to draw pictures of molecules, in which he enclosed the symbols for atoms in circles, and used broken lines to connect the atoms together in a way that satisfied each atom's valence.

The year 1873, by many accounts, was a seminal point in the history of the development of the concept of the "molecule". In this year, the renowned Scottish physicist James Clerk Maxwell published his famous thirteen page article 'Molecules' in the September issue of *Nature*.[14] In the opening section to this article, Maxwell clearly states:

> An atom is a body which cannot be cut in two; a **molecule** is the smallest possible portion of a particular substance.

After speaking about the atomic theory of Democritus, Maxwell goes on to tell us that the word 'molecule' is a modern word. He states, "it does not occur in *Johnson's Dictionary*. The ideas it embodies are those belonging to modern chemistry." We are told that an 'atom' is a material point, invested and surrounded by 'potential forces' and that when 'flying molecules' strike against a solid body in constant succession it causes what is called pressure of air and other gases. At this point, however, Maxwell notes that no one has ever seen or handled a molecule.

In 1874, Jacobus Henricus van 't Hoff and Joseph Achille Le Bel independently proposed that the phenomenon of optical activity could be explained by assuming that the chemical bonds between carbon atoms and their neighbors were directed towards the corners of a regular tetrahedron. This led to a better understanding of the three-dimensional nature of molecules.

Emil Fischer developed the Fischer projection technique for viewing 3-D molecules on a 2-D sheet of paper:

In 1898, Ludwig Boltzmann, in his *Lectures on Gas Theory*, used the theory of valence to explain the phenomenon of gas phase molecular dissociation, and in doing so drew one of the first rudimentary yet detailed atomic orbital overlap drawings. Noting first the known fact that molecular iodine vapor dissociates into atoms at higher temperatures, Boltzmann states that we must explain the existence of molecules composed of two atoms, the "double atom" as Boltzmann calls it, by an attractive force acting between the two atoms. Boltzmann states that this chemical attraction, owing to certain facts of chemical valence, must be associated with a relatively small region on the surface of the atom called the *sensitive region*.

Boltzmann states that this "sensitive region" will lie on the surface of the atom, or may partially lie inside the atom, and will firmly be connected to it. Specifically, he states "only when two atoms are situated so that their sensitive regions are in contact, or partly overlap, will there be a chemical attraction between them. We then say that they are chemically bound to each other." This picture is detailed below, showing the *α-sensitive region* of atom-A overlapping with the *β-sensitive region* of atom-B:[15]

4.4 20th century

In the early 20th century, the American chemist Gilbert N. Lewis began to use dots in lecture, while teaching undergraduates at Harvard, to represent the electrons around atoms. His students favored these drawings, which stimulated

Benzene molecule with alternating double bonds

him in this direction. From these lectures, Lewis noted that elements with a certain number of electrons seemed to have a special stability. This phenomenon was pointed out by the German chemist Richard Abegg in 1904, to which Lewis referred to as "Abegg's law of valence" (now generally known as Abegg's rule). To Lewis it appeared that once a core of eight electrons has formed around a nucleus, the layer is filled, and a new layer is started. Lewis also noted that various ions with eight electrons also seemed to have a special stability. On these views, he proposed the rule of eight or octet rule: *Ions or atoms with a filled layer of eight electrons have a special stability.*[16]

Hofmann's 1865 stick-and-ball model of methane CH_4.

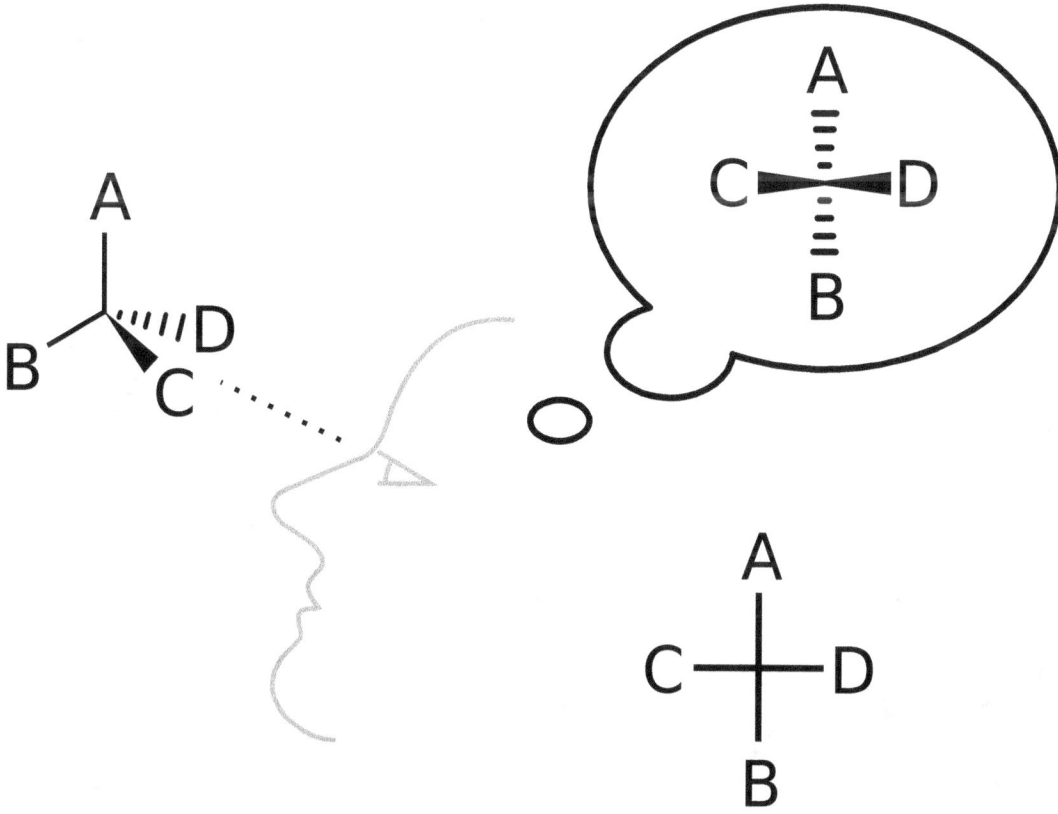

Moreover, noting that a cube has eight corners Lewis envisioned an atom as having eight sides available for electrons, like the corner of a cube. Subsequently, in 1902 he devised a conception in which cubic atoms can bond on their sides to form cubic-structured molecules.

In other words, electron-pair bonds are formed when two atoms share an edge, as in structure **C** below. This results in the sharing of two electrons. Similarly, charged ionic-bonds are formed by the transfer of an electron from one cube to another, without sharing an edge **A**. An intermediate state **B** where only one corner is shared was also postulated by Lewis.

Hence, double bonds are formed by sharing a face between two cubic atoms. This results in the sharing of four electrons.

In 1913, while working as the chair of the department of chemistry at the University of California, Berkeley, Lewis read a preliminary outline of paper by an English graduate student, Alfred Lauck Parson, who was visiting Berkeley for a year. In this paper, Parson suggested that the electron is not merely an electric charge but is also a small magnet (or "magneton" as he called it) and furthermore that a chemical bond results from two electrons being shared between two atoms.[17] This, according to Lewis, meant that bonding occurred when two electrons formed a shared edge between two complete cubes.

On these views, in his famous 1916 article *The Atom and the Molecule*, Lewis introduced the "Lewis structure" to represent atoms and molecules, where dots represent electrons and lines represent covalent bonds. In this article, he developed the concept of the electron-pair bond, in which two atoms may share one to six electrons, thus forming the single electron bond, a single bond, a double bond, or a triple bond.

In Lewis' own words:

> An electron may form a part of the shell of two different atoms and cannot be said to belong to either one exclusively.

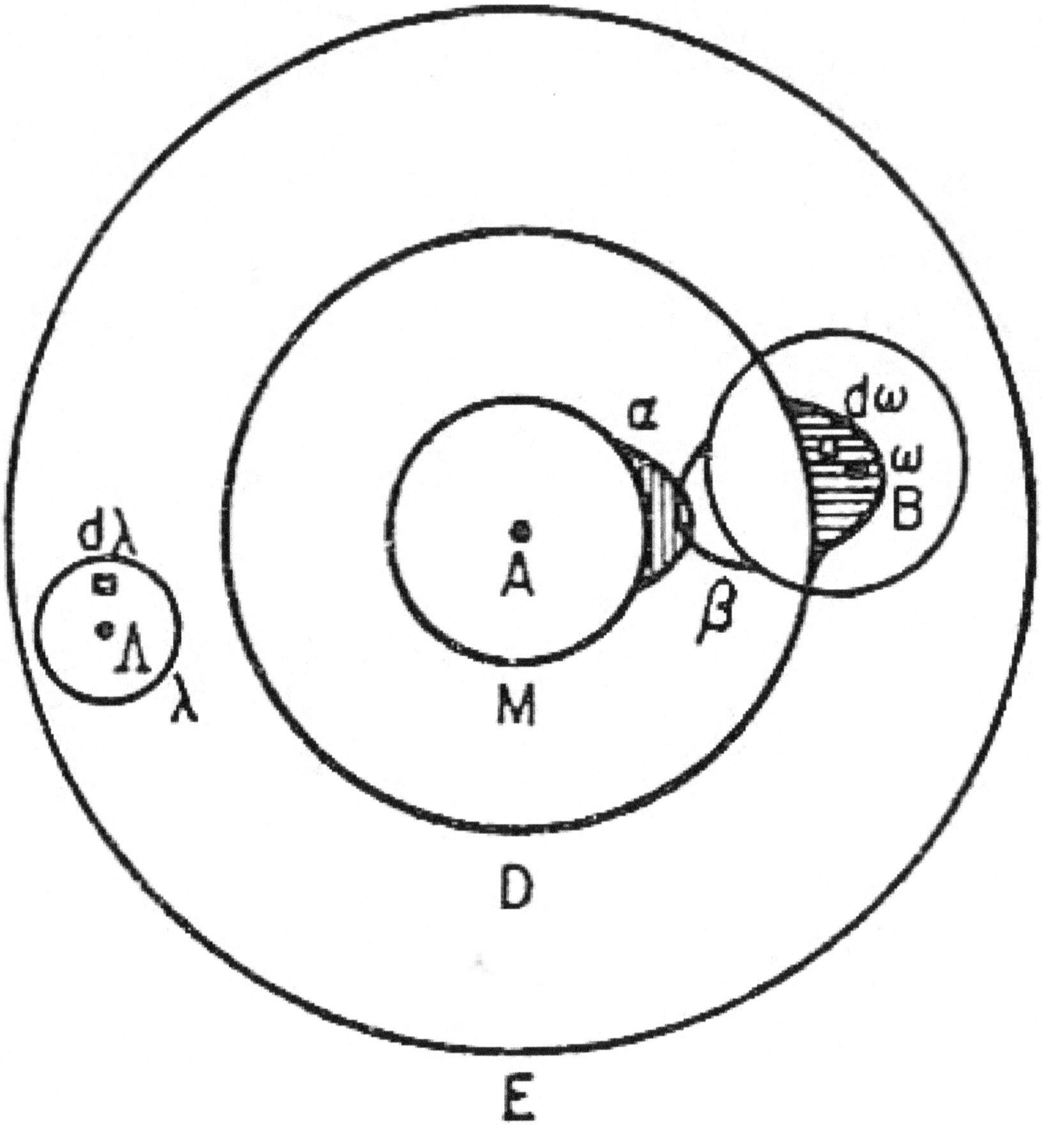

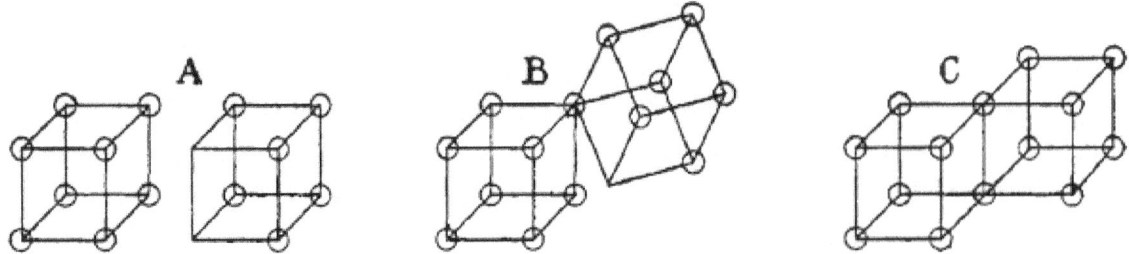

Boltzmann's 1898 I_2 molecule diagram showing atomic "sensitive region" (α, β) overlap.

*Lewis cubic-atoms bonding to form **cubic-molecules***

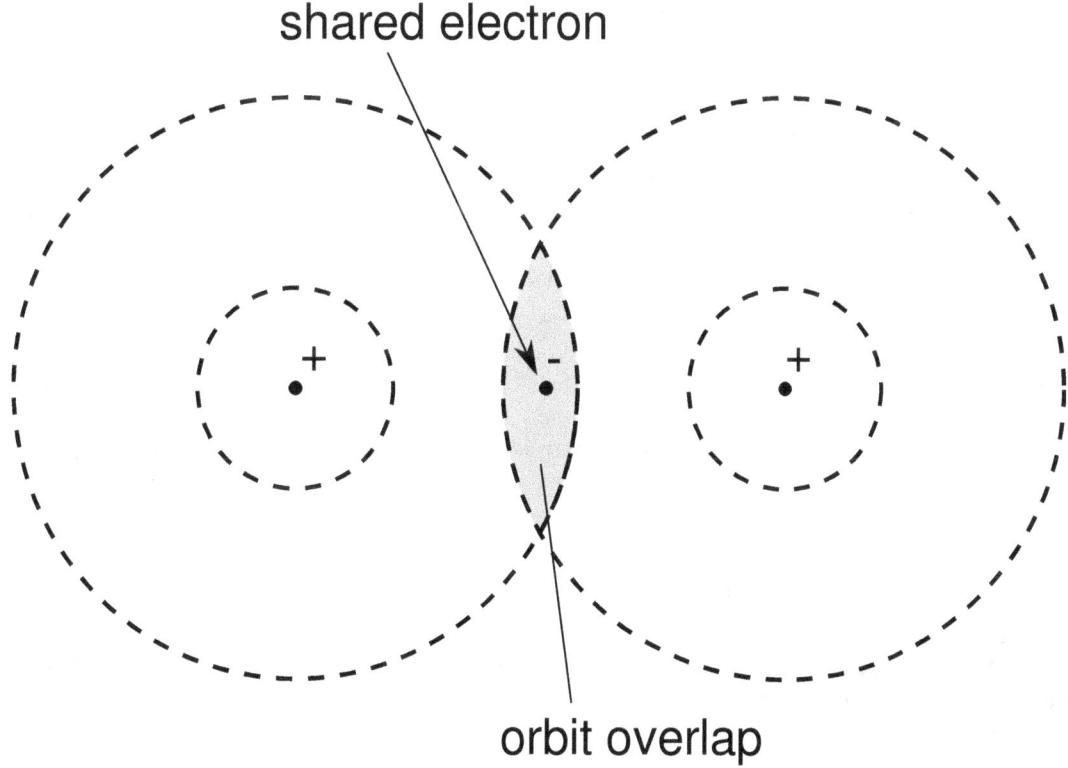

Lewis-type Chemical bond

Moreover, he proposed that an atom tended to form an ion by gaining or losing the number of electrons needed to complete a cube. Thus, Lewis structures show each atom in the structure of the molecule using its chemical symbol. Lines are drawn between atoms that are bonded to one another; occasionally, pairs of dots are used instead of lines. Excess electrons that form lone pairs are represented as pair of dots, and are placed next to the atoms on which they reside:

Lewis dot structures of the Nitrite-ion

To summarize his views on his new bonding model, Lewis states:[18]

> Two atoms may conform to the rule of eight, or the octet rule, not only by the transfer of electrons from one atom to another, but also by sharing one or more pairs of electrons...Two electrons thus coupled together, when lying between two atomic centers, and held jointly in the shells of the two atoms, I have considered to be the chemical bond. We thus have a concrete picture of that physical entity, that "hook and eye" which is part of the creed of the organic chemist.

The following year, in 1917, an unknown American undergraduate chemical engineer named Linus Pauling was learning the Dalton hook-and-eye bonding method at the Oregon Agricultural College, which was the vogue description of bonds between atoms at the time. Each atom had a certain number of hooks that allowed it to attach to other atoms, and a certain number of eyes that allowed other atoms to attach to it. A chemical bond resulted when a hook and eye connected. Pauling, however, wasn't satisfied with this archaic method and looked to the newly emerging field of quantum physics for a new method.

In 1927, the physicists Fritz London and Walter Heitler applied the new quantum mechanics to the deal with the saturable, nondynamic forces of attraction and repulsion, i.e., exchange forces, of the hydrogen molecule. Their valence bond treatment of this problem, in their joint paper,[19] was a landmark in that it brought chemistry under quantum mechanics. Their work was an influence on Pauling, who had just received his doctorate and visited Heitler and London in Zürich on a Guggenheim Fellowship.

Subsequently, in 1931, building on the work of Heitler and London and on theories found in Lewis' famous article, Pauling published his ground-breaking article "The Nature of the Chemical Bond"[20] (see: manuscript) in which he used quantum mechanics to calculate properties and structures of molecules, such as angles between bonds and rotation about bonds. On these concepts, Pauling developed hybridization theory to account for bonds in molecules such as CH_4, in which four sp^3 hybridised orbitals are overlapped by hydrogen's *1s* orbital, yielding four sigma (σ) bonds. The four bonds are of the same length and strength, which yields a molecular structure as shown below:

Owing to these exceptional theories, Pauling won the 1954 Nobel Prize in Chemistry. Notably he has been the only person to ever win two unshared Nobel prizes, winning the Nobel Peace Prize in 1963.

In 1926, French physicist Jean Perrin received the Nobel Prize in physics for proving, conclusively, the existence of molecules. He did this by calculating Avogadro's number using three different methods, all involving liquid phase systems. First, he used a gamboge soap-like emulsion, second by doing experimental work on Brownian motion, and third by confirming Einstein's theory of particle rotation in the liquid phase.[21]

In 1937, chemist K.L. Wolf introduced the concept of supermolecules (*Übermoleküle*) to describe hydrogen bonding in acetic acid dimers. This would eventually lead to the area of supermolecular chemistry, which is the study of non-covalent bonding.

In 1951, physicist Erwin Wilhelm Müller invents the field ion microscope and is the first to see atoms, e.g. bonded atomic arrangements at the tip of a metal point.

In 1999, researchers from the University of Vienna reported results from experiments on wave-particle duality for C_{60} molecules.[22] The data published by Zeilinger et al. were consistent with de Broglie wave interference for C_{60} molecules. This experiment was noted for extending the applicability of wave–particle duality by about one order of magnitude in the macroscopic direction.[23]

In 2009, researchers from IBM managed to take the first picture of a real molecule.[24] Using an atomic force microscope every single atom and bond of a pentacene molecule could be imaged.

4.5 See also

- History of chemistry

- History of quantum mechanics

- History of thermodynamics

- History of molecular biology

- Kinetic theory

- Atomic theory

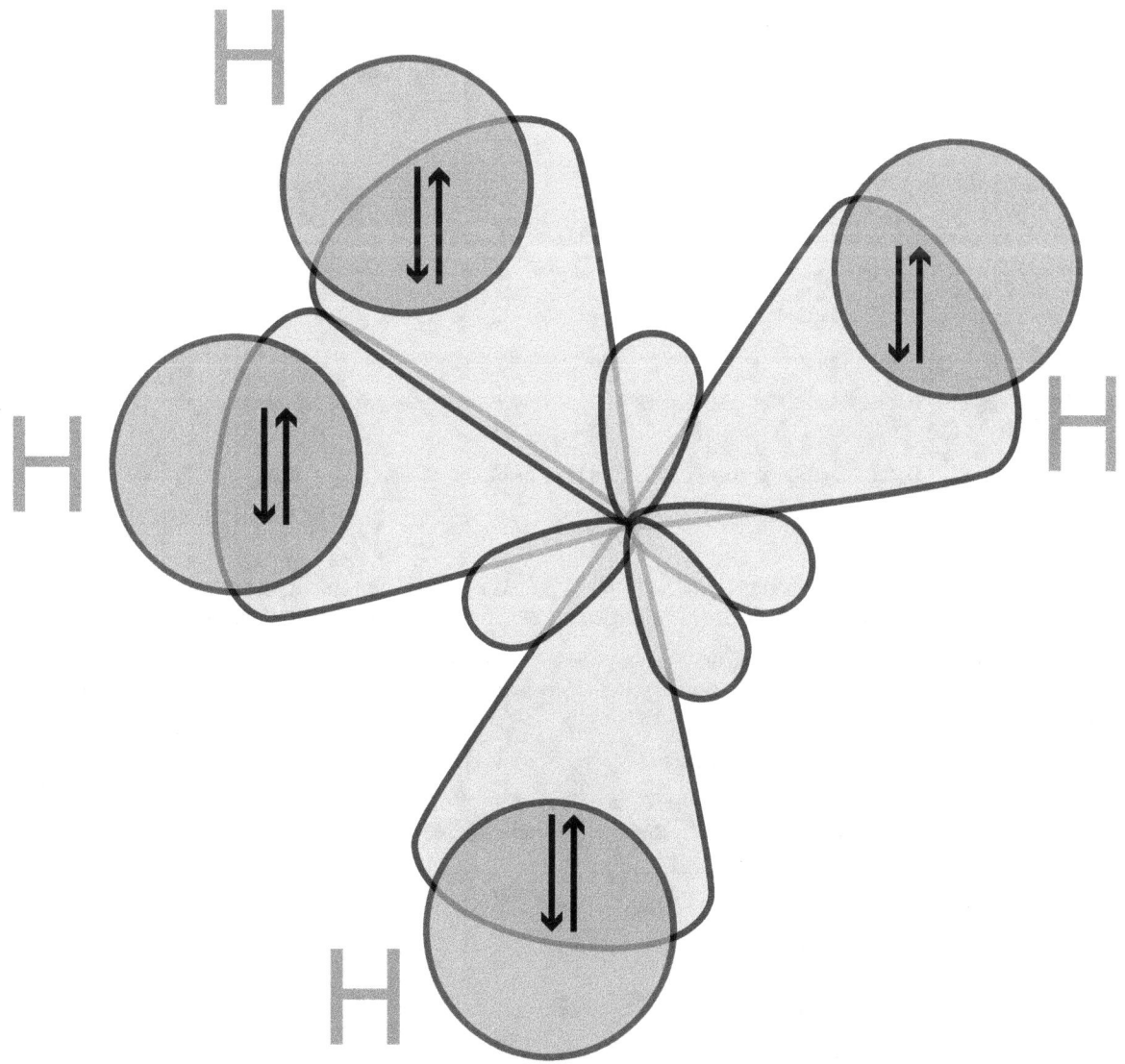

A schematic presentation of hybrid orbitals overlapping hydrogens' s orbitals

4.6   References

[1] Russell, Bertrand (2007). *A History of Western Philosophy*. Simon & Schuster. p. 41. ISBN 978-1-4165-5477-6.

[2] Russell, Bertrand (2007). *A History of Western Philosophy*. Simon & Schuster. p. 145. ISBN 978-1-4165-5477-6.

[3] Pfeffer, Jeremy, I.; Nir, Shlomo (2001). *Modern Physics: An Introduction Text*. World Scientific Publishing Company. p. 183. ISBN 1-86094-250-4.

[4] See *testimonia* DK 68 A 80, DK 68 A 37 and DK 68 A 43. See also Cassirer, Ernst (1953). *An Essay on Man: an Introduction to the Philosophy of Human Culture*. Doubleday & Co. p. 214. ISBN 0-300-00034-0. ASIN B0007EK5MM.

[5] Leicester, Henry, M. (1956). *The Historical Background of Chemistry*. John Wiley & Sons. p. 112. ISBN 0-486-61053-5.

[6] (a) Isaac Newton, (1704). Opticks. (pg. 389). New York: Dover.
    (b) Bernard, Pullman; Reisinger, Axel, R. (2001). *The Atom in the History of Human Thought*. Oxford University Press. p. 139. ISBN 0-19-515040-6.

[7] Lemery, Nicolas. (1680). An Appendix to a Course of Chymistry. London, pgs 14-15.

[8] Avogadro, Amedeo (1811). "Masses of the Elementary Molecules of Bodies". *Journal de Physique* **73**: 58–76.

[9] Seymour H. Mauskopf (1969). "The Atomic Structural Theories of Ampère and Gaudin: Molecular Speculation and Avogadro's Hypothesis". *Isis* **60** (1): 61–74. doi:10.1086/350449. JSTOR 229022.

[10] Chemical Bonding Concepts – Oklahoma State University

[11] Couper's bond line drawings (1858) – Chemical Achievers

[12] Bader, A. & Parker, L. (2001). "Joseph Loschmidt", *Physics Today*, Mar.

[13] Ollis, W. D. (1972). "Models and molecules". *Proceedings of the Royal Institution of Great Britain* **45**: 1–31.

[14] Maxwell, James Clerk, "Molecules". *Nature*, September, 1873.

[15] Boltzmann, Ludwig (1898). *Lectures on Gas Theory.* Dover (reprint). ISBN 0-486-68455-5.

[16] Cobb, Cathy (1995). *Creations of Fire - Chemistry's Lively History From Alchemy to the Atomic Age.* Perseus Publishing. ISBN 0-7382-0594-X.

[17] Parson, A.L. (1915). "A Magneton Theory of the Structure of the Atom". *Smithsonian Publication* 2371, Washington.

[18] "Valence and The Structure of Atoms and Molecules", G. N. Lewis, American Chemical Society Monograph Series, page 79 and 81.

[19] Heitler, Walter; London, Fritz (1927). "Wechselwirkung neutraler Atome und homöopolare Bindung nach der Quanten-mechanik". *Zeitschrift für Physik* **44**: 455–472. Bibcode:1927ZPhy...44..455H. doi:10.1007/BF01397394.

[20] Pauling, Linus (1931). "The nature of the chemical bond. Application of results obtained from the quantum mechanics and from a theory of paramagnetic susceptibility to the structure of molecules". *J. Am. Chem. Soc.* **53**: 1367–1400. doi:10.1021/ja01355a027.

[21] Perrin, Jean, B. (1926). Discontinuous Structure of Matter, Nobel Lecture, December 11.

[22] Arndt, M.; O. Nairz; J. Voss-Andreae; C. Keller; G. van der Zouw; A. Zeilinger (14 October 1999). "Wave-particle duality of C60 molecules". *Nature* **401** (6754): 680–682. Bibcode:1999Natur.401..680A. doi:10.1038/44348. PMID 18494170.

[23] Rae, A. I. M. (14 October 1999). "Quantum physics: Waves, particles and fullerenes". *Nature* **401** (6754): 651–653. BibcR. doi:10.1038/44294.

[24] Single molecule's stunning image.

4.7 Further reading

- Partington, J.R. (1989). *A Short History of Chemistry.* Dover Publications, Inc. ISBN 0-486-65977-1.

- Atkins, Peter (2003). *Atkins' Molecules, 2nd Ed.* Cambridge University Press. ISBN 0-521-53536-0.

- Sargent, Ted (2006). *The Dance of Molecules - How Nanotechnology is Changing our Lives.* Thunder's Mouth Press. ISBN 1-56025-809-8.

- Scerri, Eric R. (2007). *The Periodic Table, Its Story and Its Significance.* Oxford University Press. ISBN 978-0-19-530573-9.

4.8 External links

- Geometric Structures of Molecules - Middlebury College

- Atoms and Molecules - McMaster University

- 3D Molecule Viewer - The Wileys Family

- Molecule of the Month - School of Chemistry, University of Bristol

- - Eric Scerri's history & philosophy of chemistry website

4.8.1   Types

- Antibody Molecule - The National Health Museum

- 15 Types of Molecules - IUPAC Definitions

4.8.2   Definitions

- Molecule Definition - Frostburg State University (Department of Chemistry)

- Definition of Molecule - IUPAC

4.8.3   Articles

- Molecules Used to Make Nano-sized Containers - TRN Newswire

- Molecular Computer Processors - HP Labs

Chapter 5

Empirical formula

This article is about analytical chemistry. For observation rather than theory, see Empirical relationship.

In chemistry, the **empirical formula** of a chemical compound is the simplest positive integer ratio of atoms present in a compound.[1] A simple example of this concept is that the empirical formula of sulfur monoxide, or SO, would simply be SO, as is the empirical formula of disulfur dioxide, S_2O_2. This means that sulfur monoxide and disulfur dioxide, both compounds of Sulfur and Oxygen will have the same empirical formula.

An empirical formula makes no mention towards arrangement or number of atoms. It is standard for a lot of ionic compounds, like $CaCl_2$, and for macromolecules, such as SiO_2.

The molecular formula on the other hand shows the number of each type of atom in a molecule, also the structural formula shows the arrangement of the molecule. It is possible for different types of compounds to have equal empirical formulas.

5.1 Examples

- Glucose (C
  6H
  12O
  6), ribose (C
  5H
  10O
  5), acetic acid (C
  2H
  4O
  2), and formaldehyde (CH
  2O) all have different molecular formulas but the same empirical formula: CH
  2O. This is the actual molecular formula for formaldehyde, but acetic acid has double the number of atoms, ribose has five times the number of atoms, and glucose has six times the number of atoms.

- The chemical compound n-hexane has the structural formula CH
  3CH
  2CH
  2CH
  2CH
  2CH
  3, which shows that it has 6 carbon atoms arranged in a chain, and 14 hydrogen atoms. Hexane's molecular formula is C
  6H

29

14, and its empirical formula is C

3H

7, showing a C:H ratio of 3:7.

5.2   Calculation

Suppose you are given a compound such as methyl acetate, a solvent commonly used in paints, inks, and adhesives. When methyl acetate was chemically analyzed, it was discovered to have 48.64% carbon (C), 8.16% hydrogen (H), and 43.20% oxygen (O). For the purposes of determining empirical formulas, we assume that we have 100 g of the compound. If this is the case, the percentages will be equal to the mass of each element in grams.

Step 1 Change each percentage to an expression of the mass of each element in grams. That is, 48.64% C becomes 48.64 g C, 8.16% H becomes 8.16 g H, and 43.20% O becomes 43.20 g O.

Step 2 Convert the amount of each element in grams to its amount in moles.

$$\left(\frac{48.64 \text{ g C}}{1} \right) \left(\frac{1 \text{ mol}}{12.01 \text{ g C}} \right) = 4.049 \text{ mol}$$

$$\left(\frac{8.16 \text{ g H}}{1} \right) \left(\frac{1 \text{ mol}}{1.008 \text{ g H}} \right) = 8.095 \text{ mol}$$

$$\left(\frac{43.20 \text{ g O}}{1} \right) \left(\frac{1 \text{ mol}}{16.00 \text{ g O}} \right) = 2.7 \text{ mol}$$

Step 3 Divide each of the found values by the smallest of these values (2.7)

$$\frac{4.049 \text{ mol}}{2.7 \text{ mol}} = 1.5$$

$$\frac{8.095 \text{ mol}}{2.7 \text{ mol}} = 3$$

$$\frac{2.7 \text{ mol}}{2.7 \text{ mol}} = 1$$

Step 4 If necessary, multiply these numbers by integers in order to get whole numbers; if an operation is done to one of the numbers, it must be done to all of them.

$1.5 \times 2 = 3$

$3 \times 2 = 6$

$1 \times 2 = 2$

Thus, the empirical formula of methyl acetate is $C_3H_6O_2$. This formula also happens to be methyl acetate's molecular formula.

5.3   References

[1] IUPAC, *Compendium of Chemical Terminology*, 2nd ed. (the "Gold Book") (1997). Online corrected version: (2006–) "Empirical formula".

Chapter 6

Chemical substance

"Chemical" redirects here. For other uses, see Chemical (disambiguation).

The term **chemical substance** refers in technical uses to any form of matter that has constant chemical composition of its component entities, such that its physical properties that can be measured are constant; however, *chemical substance* can also have a broader meaning that encompasses regulatory and common uses that expands the former, where it can mean *any* substance produced by, used for, or related to chemistry or chemical operations or production.

Chemical substance, when used according to its technical meaning—used by specialists, generally chemists—derives from such sources as the Compendium of Chemical Terminology ("Gold Book") of the International Union of Pure and Applied Chemistry (IUPAC). As a technical term, *chemical substance* refers to any form of matter—solid, liquid, or gas—that has constant chemical composition of its component atoms, molecules, or other entities, that results in physical properties (e.g., melting point, refractive index, density, etc.) that can be measured to characterize it.

In this narrow technical definition, *chemical substances* fall into several clear subcategories: *they can be elemental materials* in which all atoms have the same atomic number[1] (e.g. metallic gold, the material diamond, the diatomic molecule H_2, or the polyatomic molecule S_8). Alternatively *they can be non-stoichiometric compounds* composed of non-integer proportions of different atoms (e.g., palladium hydride, PdHx, x ≠ 1,2,3...); or *they can be chemical compounds*, pure matter consisting of two or more different chemical elements in fixed proportion, however simple or complex—e.g. ammonia gas from a cylinder, pure water from a still, paracetemol before going into a tablet, sodium chloride and sucrose as a components in table salt and sugar, lead(II) sulfate before it goes into batteries, drugs whether man-made (e.g. aripiprazole) or isolates from natural sources (e.g., THC), or a humanized antibody (e.g., adalimumab) or even a BRCA1 gene or protein (if its structure is fully determined, including sequence, covalent modifications, and counter ions). In addition, when the type of chemical entity composing the substance is a molecule rather than an atom, the definition can include mixtures of defined, constant composition—composed to specification, with precise proportions of chemical ingredients—if such a mixture presents consistent, measurable physical properties (e.g., formulated motor oils, of given viscosity, flash point, etc.). Hence, these four subcategories are formal subsets of the category of *chemical substances* (by this technical definition), rather than being interchangeable as synonyms, as sometimes might appear to be the case in non-technical (and even occasional technical) writing venues.

However, *chemical substance* can also more broadly connote—in venues ranging from EPA documents, to databases exhibiting flexibility for sake of expedience, to common parlance—any substance produced by, used for, or related to chemistry or chemical operations or production. For instance, the EPA definition of this term is based on text from a legislative action, and includes any particular molecular identity, organic or inorganic, alone or in combination, from nature or artificial chemical reaction; this definition, therefore, falls outside the IUPAC definition, and within the general meaning of a "substance" that is "chemical."

As noted, chemical substances exist as solids, liquids, or gases, and may change between these phases of matter with changes in temperature or pressure. As the title term always refers to matter, all forms of energy (heat, light, etc.) are not chemical substances. Some types of chemical substances can generally be thought of as pure (as defined operationally, for the substance, in a particular use); chemical substance classes that can be though of in terms of purity include chemical elements, chemical compounds, and non-stoichiometric compounds (examples listed above). Other types of chemical substances that are best understood as being well-defined in composition, without reference to purity, are mixtures of defined composition that present consistent, measurable physical properties. Finally, the most loose definition in use, of any material that is chemical in origin or association, makes no claim for purity or defined composition; hence, the title term also appears, e.g., to refer to spills of chemical compounds, preparations, or mixtures of unknown composition, when occurring in the field.

6.1   Definition

Colors of a single chemical (Nile red) in different solvents, under visible and UV light, showing how the chemical interacts dynamically with its solvent environment.

Into the second decade of the new millennium, the combined title term is left undefined in major dictionaries such as Merriam-Webster and Oxford; however, *chemical substance* is both defined by the International Union of Pure and Applied Chemistry, as a technical chemistry term (see below), and it can be understood based on its more general definition as a material "substance", with the modifier "chemical". The former case, the technical, IUPAC use of *chemical substance,* is clearly a broadening of other more specific chemical categories (e.g., like *chemical compound*), but also a narrowing of the general meaning of any "substance" that can be identified as "chemical", e.g., "there was a chemical substance spilled on the floor."

The formal IUPAC definition is that a *chemical substance* is any material presenting itself that one can conclude is made up of "matter of constant composition," where the constant composition can be of any chemical component—entities such as "molecules, formula units, atoms," etc.—and where the result of the constancy is that the substance presents "[p]hysical properties such as density, refractive index, electric conductivity, melting point etc." that can be measured to characterize it.[2] This formal definition is the basis of other definitions used by particular groups or agencies—e.g., the EPA, see below,—or publications (e.g., encyclopedias[3]). Critically, by this definition, *chemical substances* include:

- all *chemical compounds*—pure matter consisting of two or more different chemical elements in fixed proportion;

- elemental materials failing the two or more atom requirement (such as in the diatomic molecule H_2, or the polyatomic molecule S_8, etc.);

- "non-stoichiometric" subset of chemical compounds whose proportions are non-integral [e.g., palladium hydride, PdHx (0.02 < x < 0.58)]; and

- mixtures of defined and constant composition that present consistent, measurable physical properties.

In the last case, formulated motor oils and other such complex mixtures qualify as they are composed to specification of precise proportions of chemical ingredients, and therefore exhibit a set of measurable physical properties (e.g., viscosity, flash point, etc.). A slight narrowing of the IUPAC definition often appears in introductory (general) chemistry textbooks, where a definition such as "any material with a definite chemical composition" is stated, omitting the consequent, derivative measurable physical properties (often appearing in later explanations).[4] In such cases, a simple example is often chosen from among the compounds, such pure water, H_2O, obscuring the more general meaning of *substance* and the more specific meaning of *compound*.

The most general meaning of *chemical substance* that appears is as a subcategory of all *substances*—any real physical matter with a tangible, presence, often composing something otherwise identified[5][6]—specifically, those substances that can, in some respect, be identified as "chemical"—anything produced by, used for, or related to chemistry or chemical operations or production.[7][8] In this regard, the EPA offers a definition based in legislative action [Section 8(b) of the *Toxic Substances Control Act*] that includes any "particular molecular identity," organic or inorganic, alone or in combination, from nature or artificial chemical reaction, and therefore appears as this more general type of meaning.[9] Hence, while *encompassing* pure and well-defined chemical elements and chemical compounds, an appropriate use of the title term remains, "the spill was a chemical substance," even though the material does not necessary derive from anything pure or composed of fixed proportions (e.g., water or sulfuric acid), nor even from anything of constant composition and reproducible physical properties (IUPAC definition, e.g., new motor oil). In short, the title term has connotations that include impure and ill-defined chemical cases.[9]

In specific non-chemistry disciplines, more nuanced versions of the foregoing definitions may apply. In geology, substances of uniform composition are called minerals, while physical mixtures (aggregates) of several minerals (different substances) are defined as rocks. Many minerals, however, mutually dissolve into solid solutions, such that a single rock is a uniform substance despite being a mixture in stoichiometric terms. Feldspars are a common example: anorthoclase is an alkali aluminium silicate, where the alkali metal is interchangeably either sodium or potassium. Finally, the term can even be used loosely within the chemical profession, when expedience dictates. For instance, the "chemical substance" index published by CAS also includes several alloys of uncertain composition.[10]

6.2 History

The term "chemical substance" became firmly established in the late eighteenth century after work by the chemist Joseph Proust on the composition of some pure chemical compounds such as basic copper carbonate.[11] He deduced that, "All samples of a compound have the same composition; that is, all samples have the same proportions, by mass, of the elements present in the compound." This is now known as the law of constant composition.[12] Later with the advancement of methods for chemical synthesis particularly in the realm of organic chemistry; the discovery of many more chemical elements and new techniques in the realm of analytical chemistry used for isolation and purification of elements and compounds from chemicals that led to the establishment of modern chemistry, the concept was defined as is found in most chemistry textbooks. However, there are some controversies regarding this definition mainly because the large number of chemical substances reported in chemistry literature need to be indexed.

Isomerism caused much consternation to early researchers, since isomers have exactly the same composition, but differ in configuration (arrangement) of the atoms. For example, there was much speculation for the chemical identity of benzene, until the correct structure was described by Friedrich August Kekulé. Likewise, the idea of stereoisomerism - that atoms have rigid three-dimensional structure and can thus form isomers that differ only in their three-dimensional arrangement - was another crucial step in understanding the concept of distinct chemical substances. For example, tartaric acid has three distinct isomers, a pair of diastereomers with one diastereomer forming two enantiomers.

6.3   Chemical elements

Native sulfur crystals. Sulfur occurs naturally as elemental sulfur, in sulfide and sulfate minerals and in hydrogen sulfide.

Main article: Chemical element
See also: List of elements

An element is a chemical substance that is made up of atoms with the same atomic number and hence cannot be broken down or transformed by a chemical reaction into a different element, though it can be transmutated into another element through a nuclear reaction. This is so, because all of the atoms in a sample of an element have the same number of protons, though they may be different isotopes, with differing numbers of neutrons.

As of 2012, there are 118 known elements, about 80 of which are stable – that is, they do not change by radioactive decay into other elements. Some elements can occur as more than a single chemical substance (allotropes). For instance, oxygen exists as both diatomic oxygen (O_2) and ozone (O_3). The majority of elements are classified as metals. These are elements with a characteristic lustre such as iron, copper, and gold. Metals typically conduct electricity and heat well, and they are malleable and ductile.[13] Around a dozen elements,[14] such as carbon, nitrogen, and oxygen, are classified as non-metals. Non-metals lack the metallic properties described above, they also have a high electronegativity and a tendency to form negative ions. Certain elements such as silicon sometimes resemble metals and sometimes resemble non-metals, and are known as metalloids.

Potassium ferricyanide is a compound of potassium, iron, carbon and nitrogen; although it contains cyanide anions, it does not release them and is nontoxic.

6.4 Chemical compounds

Main article: Chemical compound
See also: List of organic compounds and List of inorganic compounds

A pure chemical compound is a chemical substance that is composed of a particular set of molecules or ions. Two or more elements combined into one substance through a chemical reaction form a chemical compound. All compounds are substances, but not all substances are compounds.

A chemical compound can be either atoms bonded together in molecules or crystals in which atoms, molecules or ions form a crystalline lattice. Compounds based primarily on carbon and hydrogen atoms are called organic compounds, and all others are called inorganic compounds. Compounds containing bonds between carbon and a metal are called organometallic compounds.

Compounds in which components share electrons are known as covalent compounds. Compounds consisting of oppositely charged ions are known as ionic compounds, or salts.

In organic chemistry, there can be more than one chemical compound with the same composition and molecular weight. Generally, these are called isomers. Isomers usually have substantially different chemical properties, may be isolated and do not spontaneously convert to each other. A common example is glucose vs. fructose. The former is an aldehyde, the latter is a ketone. Their interconversion requires either enzymatic or acid-base catalysis. However, there are also tautomers, where isomerization occurs spontaneously, such that a pure substance cannot be isolated into its tautomers. A common example is glucose, which has open-chain and ring forms. One cannot manufacture pure open-chain glucose because glucose spontaneously cyclizes to the hemiacetal form. Materials may also comprise other entities such as polymers. These may be inorganic or organic and sometimes a combination of inorganic and organic.

6.5   Substances versus mixtures

Cranberry glass, while it looks homogeneous, is a mixture *consisting of glass and gold colloidal particles of ca. 40 nm diameter, which give it a red color.*

Main article: Mixture

All matter consists of various elements and chemical compounds, but these are often intimately mixed together. Mixtures contain more than one chemical substance, and they do not have a fixed composition. In principle, they can be separated into the component substances by purely mechanical processes. Butter, soil and wood are common examples of mixtures.

Grey iron metal and yellow sulfur are both chemical elements, and they can be mixed together in any ratio to form a yellow-grey mixture. No chemical process occurs, and the material can be identified as a mixture by the fact that the sulfur and the iron can be separated by a mechanical process, such as using a magnet to attract the iron away from the sulfur.

In contrast, if iron and sulfur are heated together in a certain ratio (1 atom of iron for each atom of sulfur, or by weight, 56 grams (1 mol) of iron to 32 grams (1 mol) of sulfur), a chemical reaction takes place and a new substance is formed, the compound iron(II) sulfide, with chemical formula FeS. The resulting compound has all the properties of a chemical substance and is not a mixture. Iron(II) sulfide has its own distinct properties such as melting point and solubility, and the two elements cannot be separated using normal mechanical processes; a magnet will be unable to recover the iron, since there is no metallic iron present in the compound.

6.6 Chemicals versus chemical substances

While the term *chemical substance* is a precise technical term that is synonymous with "chemical" for professional chemists, the meaning of the word *chemical* varies for non-chemists within the English speaking world or those using English. For industries, government and society in general in some countries,[15] the word *chemical* includes a wider class of substances that contain many mixtures of such chemical substances, often finding application in many vocations.[16] In countries that require a list of ingredients in products, the "chemicals" listed would be equated with "chemical substances".[17]

Within the chemical industry, manufactured "chemicals" are chemical substances, which can be classified by production volume into bulk chemicals, fine chemicals and chemicals found in research only:

- Bulk chemicals are produced in very large quantities, usually with highly optimized continuous processes and to a relatively low price.

- Fine chemicals are produced at a high cost in small quantities for special low-volume applications such as biocides, pharmaceuticals and speciality chemicals for technical applications.

- Research chemicals are produced individually for research, such as when searching for synthetic routes or screening substances for pharmaceutical activity. In effect, their price per gram is very high, although they are not sold.

The cause of the difference in production volume is the complexity of the molecular structure of the chemical. Bulk chemicals are usually much less complex. While fine chemicals may be more complex, many of them are simple enough to be sold as "building blocks" in the synthesis of more complex molecules targeted for single use, as named above. The *production* of a chemical includes not only its synthesis but also its purification to eliminate by-products and impurities involved in the synthesis. The last step in production should be the analysis of batch lots of chemicals in order to identify and quantify the percentages of impurities for the buyer of the chemicals. The required purity and analysis depends on the application, but higher tolerance of impurities is usually expected in the production of bulk chemicals. Thus, the user of the chemical in the US might choose between the bulk or "technical grade" with higher amounts of impurities or a much purer "pharmaceutical grade" (labeled "USP", United States Pharmacopeia).

6.7 Naming and indexing

Every chemical substance has one or more systematic names, usually named according to the IUPAC rules for naming. An alternative system is used by the Chemical Abstracts Service (CAS).

Many compounds are also known by their more common, simpler names, many of which predate the systematic name. For example, the long-known sugar glucose is now systematically named 6-(hydroxymethyl)oxane-2,3,4,5-tetrol. Natural products and pharmaceuticals are also given simpler names, for example the mild pain-killer Naproxen is the more common name for the chemical compound (S)−6-methoxy-α-methyl-2-naphthaleneacetic acid.

Chemists frequently refer to chemical compounds using chemical formulae or molecular structure of the compound. There has been a phenomenal growth in the number of chemical compounds being synthesized (or isolated), and then reported in the scientific literature by professional chemists around the world.[18] An enormous number of chemical compounds are possible through the chemical combination of the known chemical elements. As of May 2011, about sixty million chemical compounds are known.[19] The names of many of these compounds are often nontrivial and hence not very easy to remember or cite accurately. Also it is difficult to keep the track of them in the literature. Several international organizations like IUPAC and CAS have initiated steps to make such tasks easier. CAS provides the abstracting services of the chemical literature, and provides a numerical identifier, known as CAS registry number to each chemical substance that has been reported in the chemical literature (such as chemistry journals and patents). This information is compiled as a database and is popularly known as the Chemical substances index. Other computer-friendly systems that have been developed for substance information, are: SMILES and the International Chemical Identifier or InChI.

6.8   Isolation, characterization, and identification

Chemical substances, depending on whether the narrow IUPAC or broad definition is being discussed, may be pure (as in the case of chemical elements and chemical compounds), or may be mixtures, well- or ill-defined (see Definitions). If a further chemical substance is desired in pure form from any type of mixture—e.g., from a natural source or from a chemical reaction mixture—the aim is isolation (purification) if individual entities from a sample that often contains numerous component chemical entities, followed by characterization of properties and structure, and then identification (if already known) or description (if new) of the purified entity.

6.9   See also

- Chemical safety signs

- IUPAC nomenclature

- Prices of elements and their compounds

6.10   Notes and references

[1] IUPAC (ed.). "chemical element". *http://iupac.org". doi:10.1351/goldbook.C01022.*

[2] IUPAC, *Compendium of Chemical Terminology*, 2nd ed. (the "Gold Book") (1997). Online corrected version: (2006–) "Chemical Substance".

[3] E.g., for the comparable definition from Dirac Delta Consultants, see: "Pure Substance – DiracDelta Science & Engineering Encyclopedia". Diracdelta.co.uk. Retrieved 2013-06-06.

[4] See, for instance, Hill, J. W.; Petrucci, R. H.; McCreary, T. W. & Perry, S.S., 2005, *General Chemistry,* 4th edn., p. 5, Upper Saddle River, NJ, USA:Pearson/Prentice Hall.

[5] "Substance". *Merriam-Webster's Collegiate® Dictionary, Eleventh Edition (online).* Springfield, MA, USA: Merriam-Webster. 2015. Retrieved 2 July 2015.

[6] "Substance". *Oxford Dictionaries (online).* Oxford, OXF, GBR: Oxford University Press. 2015. Retrieved 2 July 2015.

[7] "Chemical". *Merriam-Webster's Collegiate® Dictionary, Eleventh Edition (online).* Springfield, MA, USA: Merriam-Webster. 2015. Retrieved 2 July 2015.

[8] "Chemical". *Oxford Dictionaries (online).* Oxford, OXF, GBR: Oxford University Press. 2015. Retrieved 2 July 2015.

[9] EPA, 2014, "Basic Information: Background" at *TSCA Chemical Substance Inventory* (online), per Section 8 (b) of the Toxic Substances Control Act, see , accessed 2 July 2015. The full definition there is:"'chemical substance' means any organic or inorganic substance of a particular molecular identity, including any combination of these substances occurring in whole or in part as a result of a chemical reaction or occurring in nature, and any element or uncombined radical. Chemicals substances on the Inventory include: organics, inorganics, polymers, and UVCBs (chemical substances of unknown or variable composition, complex reaction products, and biological materials)."

[10] Appendix IV: Chemical Substance Index Names

[11] Hill, J. W.; Petrucci, R. H.; McCreary, T. W.; Perry, S. S. *General Chemistry*, 4th ed., p37, Pearson Prentice Hall, Upper Saddle River, New Jersey, 2005.

[12] Law of Definite Proportions

[13] Hill, J. W.; Petrucci, R. H.; McCreary, T. W.; Perry, S. S. *General Chemistry*, 4th ed., pp 45–46, Pearson Prentice Hall, Upper Saddle River, New Jersey, 2005.

[14] The boundary between metalloids and non-metals is imprecise, as explained in the previous reference.

[15] "What is a chemical". Nicnas.gov.au. 2005-06-01. Retrieved 2013-06-06.

[16] "BfR – Chemicals". Bfr.bund.de. 1980-09-18. Retrieved 2013-06-06.

[17] There is only one definition for "chemical", that of a substance, in the US Unabridged Edition of the Random House Dictionary of the English Language, New York, 1966.

[18] Joachim Schummer. "Coping with the Growth of Chemical Knowledge: Challenges for Chemistry Documentation, Education, and Working Chemists". Rz.uni-karlsruhe.de. Retrieved 2013-06-06.

[19] "Chemical Abstracts substance count". Cas.org. Retrieved 2013-06-06.

6.11 Further reading

- N. H. Ray, 1979, *Inorganic polymers,* New York, NY, USA:John Wiley and Sons.

6.12 External links

- eChemPortal substance and property search

- Chemical Reactions

Chapter 7

Chemical element

A **chemical element** (or **element**) is a chemical substance consisting of atoms having the same number of protons in their atomic nuclei (i.e. the same atomic number, Z).[1] There are 118 elements that have been identified, of which the first 98 occur naturally on Earth with the remaining 20 being Synthetic elements. There are 80 elements that have at least one stable isotope and 38 that have only radioactive isotopes, which decay over time into other elements. Iron is the most abundant element (by mass) making up the Earth, while oxygen is the most common element in the crust of the earth.[2]

Chemical elements constitute approximately 15% of the matter in the universe: the remainder is dark matter, the composition of it is unknown, but it is not composed of chemical elements.[3] The two lightest elements, hydrogen and helium were mostly formed in the Big Bang and are the most common elements in the universe. The next three elements (lithium, beryllium and boron) were formed mostly by cosmic ray spallation, and are thus more rare than those that follow. Formation of elements with from six to twenty six protons occurred and continues to occur in main sequence stars via stellar nucleosynthesis. The high abundance of oxygen, silicon, and iron on Earth reflects their common production in such stars. Elements with greater than twenty six protons are formed by supernova nucleosynthesis in supernovae, which, when they explode, blast these elements far into space as planetary nebulae, where they may become incorporated into planets when they are formed.[4]

When different elements are chemically combined, with the atoms held together by chemical bonds, they form chemical compounds. Only a minority of elements are found uncombined as relatively pure minerals. Among the more common of such "native elements" are copper, silver, gold, carbon (as coal, graphite, or diamonds), and sulfur. All but a few of the most inert elements, such as noble gases and noble metals, are usually found on Earth in chemically combined form, as chemical compounds. While about 32 of the chemical elements occur on Earth in native uncombined forms, most of these occur as mixtures. For example, atmospheric air is primarily a mixture of nitrogen, oxygen, and argon, and native solid elements occur in alloys, such as that of iron and nickel.

The history of the discovery and use of the elements began with primitive human societies that found native elements like carbon, sulfur, copper and gold. Later civilizations extracted elemental copper, tin, lead and iron from their ores by smelting, using charcoal. Alchemists and chemists subsequently identified many more, with almost all of the naturally-occurring elements becoming known by 1900.

The properties of the chemical elements are summarized on the periodic table, which organizes the elements by increasing atomic number into rows ("periods") in which the columns ("groups") share recurring ("periodic") physical and chemical properties. Save for unstable radioactive elements with short half-lives, all of the elements are available industrially, most of them in high degrees of purity.

7.1 Description

The lightest chemical elements are hydrogen and helium, both created by Big Bang nucleosynthesis during the first 20 minutes of the universe[5] in a ratio of around 3:1 by mass (or 12:1 by number of atoms).[6][7] Almost all other elements found in nature were made by various natural methods of nucleosynthesis.[8] On Earth, small amounts of new atoms are naturally produced in nucleogenic reactions, or in cosmogenic processes, such as cosmic ray spallation. New atoms are also naturally produced on Earth as radiogenic daughter isotopes of ongoing radioactive decay processes such as alpha decay, beta decay, spontaneous fission, cluster decay, and other rarer modes of decay.

Of the 98 naturally occurring elements, those with atomic numbers 1 through 82 each have at least one stable isotope, (except for technetium, element 43 and promethium, element 61, which have no stable isotopes). Isotopes considered stable are those for which no radioactive decay has yet been observed. Elements with atomic numbers 83 through 98 are unstable to the point that radioactive decay of all isotopes can be detected. Some of these elements, notably bismuth (atomic number 83), thorium (atomic number 90), uranium (atomic number 92) and plutonium (atomic number 94), have one or more isotopes with half-lives long enough to survive as remnants of the explosive stellar nucleosynthesis that produced the heavy elements before the formation of our solar system. For example, at over 1.9×10^{19} years, over a billion times longer than the current estimated age of the universe, bismuth-209 (atomic number 83) has the longest known alpha decay half-life of any naturally occurring element.[9][10] The very heaviest elements (those beyond californium, atomic number 98) undergo radioactive decay with half-lives so short that they do not occur in nature and must be synthesized.

As of 2010, there are 118 known elements (in this context, "known" means observed well enough, even from just a few decay products, to have been differentiated from other elements).[11][12] Of these 118 elements, 98 occur naturally on Earth.[13] Ten of these occur in extreme trace quantities: technetium, atomic number 43; promethium, number 61; astatine, number 85; francium, number 87; neptunium, number 93; plutonium, number 94; americium, number 95; curium, number 96; berkelium, number 97; and californium, number 98. These 98 elements have been detected in the universe at large, in the spectra of stars and also supernovae, where short-lived radioactive elements are newly being made. The first 98 elements have been detected directly on Earth as primordial nuclides present from the formation of the solar system, or as naturally-occurring fission or transmutation products of uranium and thorium.

The remaining 20 heavier elements, not found today either on Earth or in astronomical spectra, have been produced artificially: these are all radioactive, with very short half-lives; if any atoms of these elements were present at the formation of Earth, they are extremely likely, to the point of certainty, to have already decayed, and if present in novae, have been in quantities too small to have been noted. Technetium was the first purportedly non-naturally occurring element synthesized, in 1937, although trace amounts of technetium have since been found in nature (and also the element may have been discovered naturally in 1925).[14] This pattern of artificial production and later natural discovery has been repeated with several other radioactive naturally-occurring rare elements.[15]

Lists of the elements are available by name, by symbol, by atomic number, by density, by melting point, and by boiling point as well as ionization energies of the elements. The nuclides of stable and radioactive elements are also available as a list of nuclides, sorted by length of half-life for those that are unstable. One of the most convenient, and certainly the most traditional presentation of the elements, is in the form of the periodic table, which groups together elements with similar chemical properties (and usually also similar electronic structures).

7.1.1 Atomic number

Main article: atomic number

The atomic number of an element is equal to the number of protons in each atom, and defines the element.[16] For example, all carbon atoms contain 6 protons in their atomic nucleus; so the atomic number of carbon is 6.[17] Carbon atoms may have different numbers of neutrons; atoms of the same element having different numbers of neutrons are known as isotopes of the element.[18]

The number of protons in the atomic nucleus also determines its electric charge, which in turn determines the number of electrons of the atom in its non-ionized state. The electrons are placed into atomic orbitals that determine the atom's various chemical properties. The number of neutrons in a nucleus usually has very little effect on an element's chemical properties (except in the case of hydrogen and deuterium). Thus, all carbon isotopes have nearly identical chemical properties because they all have six protons and six electrons, even though carbon atoms may, for example, have 6 or 8 neutrons. That is why the atomic number, rather than mass number or atomic weight, is considered the identifying characteristic of a chemical element.

The symbol for atomic number is Z.

7.1.2   Isotopes

Main articles: Isotope, Stable isotope and List of nuclides

Isotopes are atoms of the same element (that is, with the same number of protons in their atomic nucleus), but having *different* numbers of neutrons. Most (66 of 94) naturally occurring elements have more than one stable isotope. Thus, for example, there are three main isotopes of carbon. All carbon atoms have 6 protons in the nucleus, but they can have either 6, 7, or 8 neutrons. Since the mass numbers of these are 12, 13 and 14 respectively, the three isotopes of carbon are known as carbon-12, carbon-13, and carbon-14, often abbreviated to ^{12}C, ^{13}C, and ^{14}C. Carbon in everyday life and in chemistry is a mixture of ^{12}C (about 98.9%), ^{13}C (about 1.1%) and about 1 atom per trillion of ^{14}C.

Except in the case of the isotopes of hydrogen (which differ greatly from each other in relative mass—enough to cause chemical effects), the isotopes of a given element are chemically nearly indistinguishable.

All of the elements have some isotopes that are radioactive (radioisotopes), although not all of these radioisotopes occur naturally. The radioisotopes typically decay into other elements upon radiating an alpha or beta particle. If an element has isotopes that are not radioactive, these are termed "stable" isotopes. All of the known stable isotopes occur naturally (see primordial isotope). The many radioisotopes that are not found in nature have been characterized after being artificially made. Certain elements have no stable isotopes and are composed *only* of radioactive isotopes: specifically the elements without any stable isotopes are technetium (atomic number 43), promethium (atomic number 61), and all observed elements with atomic numbers greater than 82.

Of the 80 elements with at least one stable isotope, 26 have only one single stable isotope. The mean number of stable isotopes for the 80 stable elements is 3.1 stable isotopes per element. The largest number of stable isotopes that occur for a single element is 10 (for tin, element 50).

7.1.3   Isotopic mass and atomic mass

Main articles: atomic mass and relative atomic mass

The mass number of an element, A, is the number of nucleons (protons and neutrons) in the atomic nucleus. Different isotopes of a given element are distinguished by their mass numbers, which are conventionally written as a superscript on the left hand side of the atomic symbol (e.g., ^{238}U). The mass number is always a simple whole number and has units of "nucleons." An example of a referral to a mass number is "magnesium-24," which is an atom with 24 nucleons (12 protons and 12 neutrons).

Whereas the mass number simply counts the total number of neutrons and protons and is thus a natural (or whole) number, the atomic mass of a single atom is a real number for the mass of a particular isotope of the element, the unit being **u**. In general, when expressed in **u** it differs in value slightly from the mass number for a given nuclide (or isotope) since the mass of the protons and neutrons is not exactly 1 **u**, since the electrons contribute a lesser share to the atomic mass as neutron number exceeds proton number, and (finally) because of the nuclear binding energy. For example, the atomic mass of chlorine-35 to five significant digits is 34.969 **u** and that of chlorine-37 is 36.966 **u**. However, the atomic mass in **u** of each isotope is quite close to its simple mass number (always within 1%). The only isotope whose atomic mass is exactly a natural number is ^{12}C, which by definition has a mass of exactly 12, because **u** is defined as 1/12 of the mass of a free neutral carbon-12 atom in the ground state.

The relative atomic mass (historically and commonly also called "atomic weight") of an element is the *average* of the atomic masses of all the chemical element's isotopes as found in a particular environment, weighted by isotopic abundance, relative to the atomic mass unit (**u**). This number may be a fraction that is *not* close to a whole number, due to the averaging process. For example, the relative atomic mass of chlorine is 35.453 **u**, which differs greatly from a whole number due to being made of an average of 76% chlorine-35 and 24% chlorine-37. Whenever a relative atomic mass value differs by more than 1% from a whole number, it is due to this averaging effect resulting from significant amounts of more than one isotope being naturally present in the sample of the element in question.

7.1.4 Chemically pure and isotopically pure

Chemists and nuclear scientists have different definitions of a *pure element*. In chemistry, a pure element means a substance whose atoms all (or in practice almost all) have the same atomic number, or number of protons. Nuclear scientists, however, define a pure element as one that consists of only one stable isotope.[19]

For example, a copper wire is 99.99% chemically pure if 99.99% of its atoms are copper, with 29 protons each. However it is not isotopically pure since ordinary copper consists of two stable isotopes, 69% ^{63}Cu and 31% ^{65}Cu, with different numbers of neutrons.

7.1.5 Allotropes

Main article: Allotropy

Atoms of chemically pure elements may bond to each other chemically in more than one way, allowing the pure element to exist in multiple structures (spatial arrangements of atoms), known as allotropes, which differ in their properties. For example, carbon can be found as diamond, which has a tetrahedral structure around each carbon atom; graphite, which has layers of carbon atoms with a hexagonal structure stacked on top of each other; graphene, which is a single layer of graphite that is very strong; fullerenes, which have nearly spherical shapes; and carbon nanotubes, which are tubes with a hexagonal structure (even these may differ from each other in electrical properties). The ability of an element to exist in one of many structural forms is known as 'allotropy'.

The standard state, also known as reference state, of an element is defined as its thermodynamically most stable state at 1 bar at a given temperature (typically at 298.15 K). In thermochemistry, an element is defined to have an enthalpy of formation of zero in its standard state. For example, the reference state for carbon is graphite, because the structure of graphite is more stable than that of the other allotropes.

7.1.6 Properties

Several kinds of descriptive categorizations can be applied broadly to the elements, including consideration of their general physical and chemical properties, their states of matter under familiar conditions, their melting and boiling points, their densities, their crystal structures as solids, and their origins.

General properties

Several terms are commonly used to characterize the general physical and chemical properties of the chemical elements. A first distinction is between metals, which readily conduct electricity, nonmetals, which do not, and a small group, (the *metalloids*), having intermediate properties and often behaving as semiconductors.

A more refined classification is often shown in colored presentations of the periodic table. This system restricts the terms "metal" and "nonmetal" to only certain of the more broadly defined metals and nonmetals, adding additional terms for certain sets of the more broadly viewed metals and nonmetals. The version of this classification used in the periodic tables presented here includes: actinides, alkali metals, alkaline earth metals, halogens, lanthanides, transition metals, post-transition metals; metalloids, noble gases, polyatomic nonmetals, diatomic nonmetals, and transition metals. In this system, the alkali metals, alkaline earth metals, and transition metals, as well as the lanthanides and the actinides, are special groups of the metals viewed in a broader sense. Similarly, the polyatomic nonmetals, diatomic nonmetals and the noble gases are nonmetals viewed in the broader sense. In some presentations, the halogens are not distinguished, with astatine identified as a metalloid and the others identified as nonmetals.

States of matter

Another commonly used basic distinction among the elements is their state of matter (phase), whether solid, liquid, or gas, at a selected standard temperature and pressure (STP). Most of the elements are solids at conventional temperatures and atmospheric pressure, while several are gases. Only bromine and mercury are liquids at 0 degrees Celsius (32 degrees Fahrenheit) and normal atmospheric pressure; caesium and gallium are solids at that temperature, but melt at 28.4 °C (83.2 °F) and 29.8 °C (85.6 °F), respectively.

Melting and boiling points

Melting and boiling points, typically expressed in degrees Celsius at a pressure of one atmosphere, are commonly used in characterizing the various elements. While known for most elements, either or both of these measurements is still undetermined for some of the radioactive elements available in only tiny quantities. Since helium remains a liquid even at absolute zero at atmospheric pressure, it has only a boiling point, and not a melting point, in conventional presentations.

Densities

Main article: Densities of the elements (data page)

The density at a selected standard temperature and pressure (STP) is frequently used in characterizing the elements. Density is often expressed in grams per cubic centimeter (g/cm^3). Since several elements are gases at commonly encountered temperatures, their densities are usually stated for their gaseous forms; when liquefied or solidified, the gaseous elements have densities similar to those of the other elements.

When an element has allotropes with different densities, one representative allotrope is typically selected in summary presentations, while densities for each allotrope can be stated where more detail is provided. For example, the three familiar allotropes of carbon (amorphous carbon, graphite, and diamond) have densities of 1.8–2.1, 2.267, and 3.515 g/cm^3, respectively.

Crystal structures

The elements studied to date as solid samples have eight kinds of crystal structures: cubic, body-centered cubic, face-centered cubic, hexagonal, monoclinic, orthorhombic, rhombohedral, and tetragonal. For some of the synthetically produced transuranic elements, available samples have been too small to determine crystal structures.

Occurrence and origin on Earth

Chemical elements may also be categorized by their origin on Earth, with the first 98 considered naturally occurring, while those with atomic numbers beyond 98 have only been produced artificially as the synthetic products of man-made nuclear reactions.

Of the 98 naturally occurring elements, 84 are considered primordial and either stable or weakly radioactive. The remaining 14 naturally occurring elements possess half lives too short for them to have been present at the beginning of the Solar System, and are therefore considered transient elements. Of these 14 transient elements, 7 (polonium, astatine, radon, francium, radium, actinium, and protactinium) are relatively common decay products of thorium, uranium, and plutonium. The remaining 7 transient elements (technetium, promethium, neptunium, americium, curium, berkelium, and californium) occur only rarely, as products of rare nuclear reaction processes involving uranium or other heavy elements.

Elements with atomic numbers 1 through 40 are all stable, while those with atomic numbers 41 through 82 (except technetium and promethium) are metastable. The half-lives of these metastable "theoretical radionuclides" are so long (at least 100 million times longer than the estimated age of the universe) that their radioactive decay has yet to be detected by experiment. Elements with atomic numbers 83 through 98 are unstable to the point that their radioactive decay can be detected. Some of these elements, notably thorium (atomic number 90) and uranium (atomic number 92), have one or more isotopes with half-lives long enough to survive as remnants of the explosive stellar nucleosynthesis that produced the heavy elements before the formation of our solar system. For example, at over 1.9×10^{19} years, over a billion times longer than the current estimated age of the universe, bismuth-209 (atomic number 83) has the longest known alpha decay half-life of any naturally occurring element.[9][10] The very heaviest elements (those beyond californium, atomic number 98) undergo radioactive decay with short half-lives and do not occur in nature.

7.1.7 The periodic table

Main article: Periodic table

The properties of the chemical elements are often summarized using the periodic table, which powerfully and elegantly organizes the elements by increasing atomic number into rows ("periods") in which the columns ("groups") share recurring ("periodic") physical and chemical properties. The current standard table contains 118 confirmed elements as of 10 April 2010.

Although earlier precursors to this presentation exist, its invention is generally credited to the Russian chemist Dmitri Mendeleev in 1869, who intended the table to illustrate recurring trends in the properties of the elements. The layout of the table has been refined and extended over time as new elements have been discovered and new theoretical models have been developed to explain chemical behavior.

Use of the periodic table is now ubiquitous within the academic discipline of chemistry, providing an extremely useful framework to classify, systematize and compare all the many different forms of chemical behavior. The table has also found wide application in physics, geology, biology, materials science, engineering, agriculture, medicine, nutrition, environmental health, and astronomy. Its principles are especially important in chemical engineering.

7.2 Nomenclature and symbols

The various chemical elements are formally identified by their unique atomic numbers, by their accepted names, and by their symbols.

7.2.1 Atomic numbers

The known elements have atomic numbers from 1 through 118, conventionally presented as Arabic numerals. Since the elements can be uniquely sequenced by atomic number, conventionally from lowest to highest (as in a periodic table), sets of elements are sometimes specified by such notation as "through", "beyond", or "from ... through", as in "through iron", "beyond uranium", or "from lanthanum through lutetium". The terms "light" and "heavy" are sometimes also used informally to indicate relative atomic numbers (not densities!), as in "lighter than carbon" or "heavier than lead", although technically the weight or mass of atoms of an element (their atomic weights or atomic masses) do not always increase monotonically with their atomic numbers.

7.2.2 Element names

Main article: List of chemical element name etymologies

The naming of various substances now known as elements precedes the atomic theory of matter, as names were given locally by various cultures to various minerals, metals, compounds, alloys, mixtures, and other materials, although at the time it was not known which chemicals were elements and which compounds. As they were identified as elements, the existing names for anciently-known elements (e.g., gold, mercury, iron) were kept in most countries. National differences emerged over the names of elements either for convenience, linguistic niceties, or nationalism. For a few illustrative examples: German speakers use "Wasserstoff" (water substance) for "hydrogen", "Sauerstoff" (acid substance) for "oxygen" and "Stickstoff" (smothering substance) for "nitrogen", while English and some romance languages use "sodium" for "natrium" and "potassium" for "kalium", and the French, Italians, Greeks, Portuguese and Poles prefer "azote/azot/azoto" (from roots meaning "no life") for "nitrogen".

For purposes of international communication and trade, the official names of the chemical elements both ancient and more recently recognized are decided by the International Union of Pure and Applied Chemistry (IUPAC), which has decided on a sort of international English language, drawing on traditional English names even when an element's chemical symbol is based on a Latin or other traditional word, for example adopting "gold" rather than "aurum" as the name for the 79th element (Au). IUPAC prefers the British spellings "aluminium" and "caesium" over the U.S. spellings "aluminum" and "cesium", and the U.S. "sulfur" over the British "sulphur". However, elements that are practical to sell in bulk in many countries often still have locally used national names, and countries whose national language does not use the Latin alphabet are likely to use the IUPAC element names.

According to IUPAC, chemical elements are not proper nouns in English; consequently, the full name of an element is not routinely capitalized in English, even if derived from a proper noun, as in californium and einsteinium. Isotope names of chemical elements are also uncapitalized if written out, *e.g.,* carbon-12 or uranium-235. Chemical element *symbols* (such as Cf for californium and Es for einsteinium), are always capitalized (see below).

In the second half of the twentieth century, physics laboratories became able to produce nuclei of chemical elements with half-lives too short for an appreciable amount of them to exist at any time. These are also named by IUPAC, which generally adopts the name chosen by the discoverer. This practice can lead to the controversial question of which research group actually discovered an element, a question that has delayed naming of elements with atomic number of 104 and higher for a considerable time. (See element naming controversy).

Precursors of such controversies involved the nationalistic namings of elements in the late 19th century. For example, *lutetium* was named in reference to Paris, France. The Germans were reluctant to relinquish naming rights to the French, often calling it *cassiopeium*. Similarly, the British discoverer of *niobium* originally named it *columbium,* in reference to the New World. It was used extensively as such by American publications prior to international standardization.

7.2.3   Chemical symbols

For listings of current chemical symbols, symbols not currently used, and other symbols that may look like chemical symbols, see Symbol (chemistry).

Specific chemical elements

Before chemistry became a science, alchemists had designed arcane symbols for both metals and common compounds. These were however used as abbreviations in diagrams or procedures; there was no concept of atoms combining to form molecules. With his advances in the atomic theory of matter, John Dalton devised his own simpler symbols, based on circles, to depict molecules.

The current system of chemical notation was invented by Berzelius. In this typographical system, chemical symbols are not mere abbreviations—though each consists of letters of the Latin alphabet. They are intended as universal symbols for people of all languages and alphabets.

The first of these symbols were intended to be fully universal. Since Latin was the common language of science at that time, they were abbreviations based on the Latin names of metals. Cu comes from Cuprum, Fe comes from Ferrum, Ag from Argentum. The symbols were not followed by a period (full stop) as with abbreviations. Later chemical elements were also assigned unique chemical symbols, based on the name of the element, but not necessarily in English. For example, sodium has the chemical symbol 'Na' after the Latin *natrium*. The same applies to "W" (wolfram) for tungsten, "Fe" (ferrum) for iron, "Hg" (hydrargyrum) for mercury, "Sn" (stannum) for tin, "K" (kalium) for potassium, "Au" (aurum) for gold, "Ag" (argentum) for silver, "Pb" (plumbum) for lead, "Cu" (cuprum) for copper, and "Sb" (stibium) for antimony.

Chemical symbols are understood internationally when element names might require translation. There have sometimes been differences in the past. For example, Germans in the past have used "J" (for the alternate name Jod) for iodine, but now use "I" and "Iod."

The first letter of a chemical symbol is always capitalized, as in the preceding examples, and the subsequent letters, if any, are always lower case (small letters). Thus, the symbols for californium or einsteinium are Cf and Es.

General chemical symbols

There are also symbols in chemical equations for groups of chemical elements, for example in comparative formulas. These are often a single capital letter, and the letters are reserved and not used for names of specific elements. For example, an "**X**" indicates a variable group (usually a halogen) in a class of compounds, while "**R**" is a radical, meaning a compound structure such as a hydrocarbon chain. The letter "**Q**" is reserved for "heat" in a chemical reaction. "**Y**" is also often used as a general chemical symbol, although it is also the symbol of yttrium. "**Z**" is also frequently used as a general variable group. "**E**" is used in organic chemistry to denote an electron-withdrawing group. "**L**" is used to represent a general ligand in inorganic and organometallic chemistry. "**M**" is also often used in place of a general metal.

At least two additional, two-letter generic chemical symbols are also in informal usage, "**Ln**" for any lanthanide element and "**An**" for any actinide element. "**Rg**" was formerly used for any rare gas element, but the group of rare gases has now been renamed noble gases and the symbol "**Rg**" has now been assigned to the element roentgenium.

Isotope symbols

Isotopes are distinguished by the atomic mass number (total protons and neutrons) for a particular isotope of an element, with this number combined with the pertinent element's symbol. IUPAC prefers that isotope symbols be written in superscript notation when practical, for example ^{12}C and ^{235}U. However, other notations, such as carbon-12 and uranium-235, or C-12 and U-235, are also used.

As a special case, the three naturally occurring isotopes of the element hydrogen are often specified as **H** for ^{1}H (protium), **D** for ^{2}H (deuterium), and **T** for ^{3}H (tritium). This convention is easier to use in chemical equations, replacing the need to write out the mass number for each atom. For example, the formula for heavy water may be written D_2O instead of $^{2}H_2O$.

7.3   Origin of the elements

Only about 4% of the total mass of the universe is made of atoms or ions, and thus represented by chemical elements. This fraction is about 15% of the total matter, with the remainder of the matter (85%) being dark matter. The nature of dark matter is unknown, but it is not composed of atoms of chemical elements because it contains no protons, neutrons, or electrons. (The remaining non-matter part of the mass of the universe is composed of the even more mysterious dark energy).

The universe's 98 naturally occurring chemical elements are thought to have been produced by at least four cosmic processes. Most of the hydrogen and helium in the universe was produced primordially in the first few minutes of the Big Bang. Three recurrently occurring later processes are thought to have produced the remaining elements. Stellar nucleosynthesis, an ongoing process, produces all elements from carbon through iron in atomic number, but little lithium, beryllium, or boron. Elements heavier in atomic number than iron, as heavy as uranium and plutonium, are produced by explosive nucleosynthesis in supernovas and other cataclysmic cosmic events. Cosmic ray spallation (fragmentation) of carbon, nitrogen, and oxygen is important to the production of lithium, beryllium and boron.

During the early phases of the Big Bang, nucleosynthesis of hydrogen nuclei resulted in the production of hydrogen-1 (protium, ^{1}H) and helium-4 (^{4}He), as well as a smaller amount of deuterium (^{2}H) and very minuscule amounts (on the order of 10^{-10}) of lithium and beryllium. Even smaller amounts of boron may have been produced in the Big Bang, since it has been observed in some very old stars, while carbon has not.[20] It is generally agreed that no heavier elements than boron were produced in the Big Bang. As a result, the primordial abundance of atoms (or ions) consisted of roughly 75% ^{1}H, 25% ^{4}He, and 0.01% deuterium, with only tiny traces of lithium, beryllium, and perhaps boron.[21] Subsequent enrichment of galactic halos occurred due to stellar nucleosynthesis and supernova nucleosynthesis.[22] However, the element abundance in intergalactic space can still closely resemble primordial conditions, unless it has been enriched by some means.

On Earth (and elsewhere), trace amounts of various elements continue to be produced from other elements as products of natural transmutation processes. These include some produced by cosmic rays or other nuclear reactions (see cosmogenic and nucleogenic nuclides), and others produced as decay products of long-lived primordial nuclides.[23] For example, trace (but detectable) amounts of carbon-14 (^{14}C) are continually produced in the atmosphere by cosmic rays impacting nitrogen atoms, and argon-40 (^{40}Ar) is continually produced by the decay of primordially occurring but unstable potassium-40 (^{40}K). Also, three primordially occurring but radioactive actinides, thorium, uranium, and plutonium, decay through a series of recurrently produced but unstable radioactive elements such as radium and radon, which are transiently present in any sample of these metals or their ores or compounds. Seven other radioactive elements, technetium, promethium, neptunium, americium, curium, berkelium, and californium, occur only incidentally in natural materials, produced as individual atoms by natural fission of the nuclei of various heavy elements or in other rare nuclear processes.

Human technology has produced various additional elements beyond these first 98, with those through atomic number 118 now known.

7.4   Abundance

Main article: Abundance of the chemical elements

The following graph (note log scale) shows the abundance of elements in our solar system. The table shows the twelve most common elements in our galaxy (estimated spectroscopically), as measured in parts per million, by mass.[24] Nearby galaxies that have evolved along similar lines have a corresponding enrichment of elements heavier than hydrogen and helium. The more distant galaxies are being viewed as they appeared in the past, so their abundances of elements appear closer to the primordial mixture. As physical laws and processes appear common throughout the visible universe, however, scientist expect that these galaxies evolved elements in similar abundance.

The abundance of elements in the Solar System is in keeping with their origin from nucleosynthesis in the Big Bang and a number of progenitor supernova stars. Very abundant hydrogen and helium are products of the Big Bang, but the next three elements are rare since they had little time to form in the Big Bang and are not made in stars (they are, however, produced in small quantities by the breakup of heavier elements in interstellar dust, as a result of impact by cosmic rays). Beginning with carbon, elements are produced in stars by buildup from alpha particles (helium nuclei), resulting in an alternatingly larger abundance of elements with even atomic numbers (these are also more stable). In general, such elements up to iron are made in large stars in the process of becoming supernovas. Iron-56 is particularly common, since it is the most stable element that can easily be made from alpha particles (being a product of decay of radioactive nickel-56, ultimately made from 14 helium nuclei). Elements heavier than iron are made in energy-absorbing processes in large stars, and their abundance in the universe (and on Earth) generally decreases with their atomic number.

The abundance of the chemical elements on **Earth** varies from air to crust to ocean, and in various types of life. The abundance of elements in Earth's crust differs from that in the Solar system (as seen in the Sun and heavy planets like Jupiter) mainly in selective loss of the very lightest elements (hydrogen and helium) and also volatile neon, carbon (as hydrocarbons), nitrogen and sulfur, as a result of solar heating in the early formation of the solar system. Oxygen, the most abundant Earth element by mass, is retained on Earth by combination with silicon. Aluminum at 8% by mass is more common in the Earth's crust than in the universe and solar system, but the composition of the far more bulky mantle, which has magnesium and iron in place of aluminum (which occurs there only at 2% of mass) more closely mirrors the elemental composition of the solar system, save for the noted loss of volatile elements to space, and loss of iron which has migrated to the Earth's core.

The composition of the human body, by contrast, more closely follows the composition of seawater—save that the human body has additional stores of carbon and nitrogen necessary to form the proteins and nucleic acids, together with phosphorus in the nucleic acids and energy transfer molecule adenosine triphosphate (ATP) that occurs in the cells of all living organisms. Certain kinds of organisms require particular additional elements, for example the magnesium in chlorophyll in green plants, the calcium in mollusc shells, or the iron in the hemoglobin in vertebrate animals' red blood cells.

7.5 History

7.5.1 Evolving definitions

The concept of an "element" as an undivisible substance has developed through three major historical phases: Classical definitions (such as those of the ancient Greeks), chemical definitions, and atomic definitions.

Classical definitions

Ancient philosophy posited a set of classical elements to explain observed patterns in nature. These *elements* originally referred to *earth*, *water*, *air* and *fire* rather than the chemical elements of modern science.

The term 'elements' (*stoicheia*) was first used by the Greek philosopher Plato in about 360 BCE in his dialogue Timaeus, which includes a discussion of the composition of inorganic and organic bodies and is a speculative treatise on chemistry. Plato believed the elements introduced a century earlier by Empedocles were composed of small polyhedral forms: tetrahedron (fire), octahedron (air), icosahedron (water), and cube (earth).[25][26]

Aristotle, c. 350 BCE, also used the term *stoicheia* and added a fifth element called aether, which formed the heavens. Aristotle defined an element as:

> Element – one of those bodies into which other bodies can decompose, and that itself is not capable of being divided into other.[27]

Chemical definitions

In 1661, Robert Boyle proposed his theory of corpuscularism which favoured the analysis of matter as constituted by irreducible units of matter (atoms) and, choosing to side with neither Aristotle's view of the four elements nor Paracelsus' view of three fundamental elements, left open the question of the number of elements.[28] The first modern list of chemical elements was given in Antoine Lavoisier's 1789 *Elements of Chemistry*, which contained thirty-three elements, including light and caloric.[29] By 1818, Jöns Jakob Berzelius had determined atomic weights for forty-five of the forty-nine then-accepted elements. Dmitri Mendeleev had sixty-six elements in his periodic table of 1869.

From Boyle until the early 20th century, an element was defined as a pure substance that could not be decomposed into any simpler substance.[28] Put another way, a chemical element cannot be transformed into other chemical elements by chemical processes. Elements during this time were generally distinguished by their atomic weights, a property measurable with fair accuracy by available analytical techniques.

Atomic definitions

The 1913 discovery by English physicist Henry Moseley that the nuclear charge is the physical basis for an atom's atomic number, further refined when the nature of protons and neutrons became appreciated, eventually led to the current definition of an element based on atomic number (number of protons per atomic nucleus). The use of atomic numbers, rather than atomic weights, to distinguish elements has greater predictive value (since these numbers are integers), and also resolves some ambiguities in the chemistry-based view due to varying properties of isotopes and allotropes within the same element. Currently, IUPAC defines an element to exist if it has isotopes with a lifetime longer than the 10^{-14} seconds it takes the nucleus to form an electronic cloud.[30]

By 1914, seventy-two elements were known, all naturally occurring.[31] The remaining naturally occurring elements were discovered or isolated in subsequent decades, and various additional elements have also been produced synthetically, with much of that work pioneered by Glenn T. Seaborg. In 1955, element 101 was discovered and named mendelevium in honor of D.I. Mendeleev, the first to arrange the elements in a periodic manner. Most recently, the synthesis of element 118 was reported in October 2006, and the synthesis of element 117 was reported in April 2010.[32]

7.5.2   Discovery and recognition of various elements

See also: Timeline of chemical elements discoveries

Ten materials familiar to various prehistoric cultures are now known to be chemical elements: Carbon, copper, gold, iron, lead, mercury, silver, sulfur, tin, and zinc. Three additional materials now accepted as elements, arsenic, antimony, and bismuth, were recognized as distinct substances prior to 1500 AD. Phosphorus, cobalt, and platinum were isolated before 1750.

Most of the remaining naturally occurring chemical elements were identified and characterized by 1900, including:

- Such now-familiar industrial materials as aluminium, silicon, nickel, chromium, magnesium, and tungsten

- Reactive metals such as lithium, sodium, potassium, and calcium

- The halogens fluorine, chlorine, bromine, and iodine

- Gases such as hydrogen, oxygen, nitrogen, helium, argon, and neon

- Most of the rare-earth elements, including cerium, lanthanum, gadolinium, and neodymium.

- The more common radioactive elements, including uranium, thorium, radium, and radon

Elements isolated or produced since 1900 include:

- The three remaining undiscovered regularly occurring stable natural elements: hafnium, lutetium, and rhenium

- Plutonium, which was first produced synthetically in 1940 by Glenn T. Seaborg, but is now also known from a few long-persisting natural occurrences

- The seven incidentally occurring natural elements (americium, berkelium, californium, curium, neptunium, promethium, and technetium), which were all first produced synthetically but later discovered in trace amounts in certaingeo logical samples

  - Three scarce decay products of uranium or thorium, (astatine, francium, and protactinium), and

- Various synthetic transuranic elements, beginning with einsteinium, fermium, mendelevium, nobelium, and lawrencium

7.5.3   Recently discovered elements

The first transuranium element (element with atomic number greater than 92) discovered was neptunium in 1940. Since 1999 claims for the discovery of new elements have been considered by the IUPAC/IUPAP Joint Working Party. As of May 2012, only the elements up to 112, copernicium, as well as element 114 Flerovium and element 116 Livermorium have been confirmed as discovered by IUPAC, while claims have been made for synthesis of elements 113, 115, 117[33] and 118. The discovery of element 112 was acknowledged in 2009, and the name 'copernicium' and the atomic symbol 'Cn' were suggested for it.[34] The name and symbol were officially endorsed by IUPAC on 19 February 2010.[35] The heaviest element that is believed to have been synthesized to date is element 118, ununoctium, on 9 October 2006, by the Flerov Laboratory of Nuclear Reactions in Dubna, Russia.[12][36] Element 117 was the latest element claimed to be discovered, in 2009.[33] IUPAC officially recognized flerovium and livermorium, elements 114 and 116, in June 2011 and approved their names in May 2012.[37]

7.6   List of the 118 known chemical elements

The following sortable table includes the 118 known chemical elements, with the names linking to the *Wikipedia* articles on each.

- **Atomic number**, **name**, and **symbol** all serve independently as unique identifiers.

- **Names** are those accepted by IUPAC; provisional names for recently produced elements not yet formally named are in parentheses.

- **Group, period,** and **block** refer to an element's position in the periodic table. Group numbers here show the currently accepted numbering; for older alternate numberings, see Group (periodic table).

- **State of matter** *(solid, liquid,* or *gas)* applies at standard temperature and pressure conditions (STP).

- **Occurrence** distinguishes naturally occurring elements, categorized as either *primordial* or *transient* (from decay), and additional *synthetic* elements that have been produced technologically, but are not known to occur naturally.

- **Description** summarizes an element's properties using the broad categories commonly presented in periodic tables: Actinide, alkali metal, alkaline earth metal, halogen, lanthanide, metal, metalloid, noble gas, non-metal, and transition metal.

7.7   See also

- Discovery of the chemical elements

- Element collecting

- Fictional element

- Goldschmidt classification

- Island of stability

- List of elements by name

- List of the elements' densities

- List of nuclides

- Periodic Systems of Small Molecules

- Prices of elements and their compounds

- Symbol (chemical element)#Symbols not currently used

- Systematic element name

- Table of nuclides

7.8   References

[1] IUPAC (ed.). "chemical element". *http://iupac.org". doi:10.1351/goldbook.C01022.*

[2] Los Alamos National Laboratory (2011). "Periodic Table of Elements: Oxygen". Los Alamos, New Mexico: Los Alamos National Security, LLC. Retrieved 7 May 2011.

[3] Oerter, Robert (2006). *The Theory of Almost Everything: The Standard Model, the Unsung Triumph of Modern Physics.* Penguin. p. 223. ISBN 978-0-452-28786-0.

[4] E. M. Burbidge, G. R. Burbidge, W. A. Fowler, F. Hoyle (1957). "Synthesis of the Elements in Stars". *Reviews of Modern Physics* **29** (4): 547–650. Bibcode:1957RvMP...29..547B. doi:10.1103/RevModPhys.29.547.

[5] See the timeline on p.10 in Oganessian, Yu. Ts.; Utyonkov, V.; Lobanov, Yu.; Abdullin, F.; Polyakov, A.; Sagaidak, R.; Shirokovsky, I.; Tsyganov, Yu. et al. (2006). "Evidence for Dark Matter" (PDF). *Physical Review C* **74** (4): 044602. Bibcode:2006PhRvC..74d4602O. doi:10.1103/PhysRevC.74.044602.

[6] lbl.gov (2005). "The Universe Adventure Hydrogen and Helium". *Lawrence Berkeley National Laboratory U.S. Department of Energy.*

[7] astro.soton.ac.uk (January 3, 2001). "Formation of the light elements". *University of Southampton.*

[8] foothill.edu (October 18, 2006). "How Stars Make Energy and New Elements" (PDF). *Foothill College.*

[9] Dumé, B (23 April 2003). "Bismuth breaks half-life record for alpha decay". *Physicsworld.com* (Bristol, England: Institute of Physics). Retrieved 14 July 2015.

[10] de Marcillac, P; Coron, N; Dambier, G; Leblanc, J; Moalic, J-P (2003). "Experimental detection of alpha-particles from the radioactive decay of natural bismuth". *Nature* **422** (6934): 876–8. Bibcode:2003Natur.422..876D. doi:10.1038/nature01541. PMID 12712201.

[11] Sanderson, K (17 October 2006). "Heaviest element made – again". Nature News. doi:10.1038/news061016-4.

[12] Schewe, P; Stein, B (17 October 200). "Elements 116 and 118 Are Discovered". *Physics News Update*. American Institute of Physics. Retrieved 19 October 2006. Check date values in: |date= (help)

[13] scienceline.ucsb.edu/. "How many elements are there in the known universe?". *University of California, Santa Cruz.*

[14] *United States Environmental Protection Agency.* "Technetium-99". epa.gov. Retrieved 26 February 2013.

[15] *Harvard–Smithsonian Center for Astrophysics.* "ORIGIN OF HEAVY ELEMENTS". cfa.harvard.edu. Retrieved 26 February 2013.

[16] "ATOMIC NUMBER AND MASS NUMBERS". ndt-ed.org. Retrieved 17 February 2013.

[17] periodic.lanl.gov. "PERIODIC TABLE OF ELEMENTS: LANL Carbon". *Los Alamos National Laboratory.*

[18] Katsuya Yamada. "Atomic mass, isotopes, and mass number." (PDF). *Los Angeles Pierce College.*

[19] "Pure element". European Nuclear Society.

[20] Wilford, JN (14 January 1992). "Hubble Observations Bring Some Surprises". *New York Times.*

[21] Wright, EL (12 September 2004). "Big Bang Nucleosynthesis". UCLA, Division of Astronomy. Retrieved 22 February 2007.

[22] Wallerstein, George; Iben, Icko; Parker, Peter; Boesgaard, Ann; Hale, Gerald; Champagne, Arthur; Barnes, Charles; Käppeler, Franz et al. (1999). "Synthesis of the elements in stars: forty years of progress" (PDF). *Reviews of Modern Physics* **69** (4): 995–1084. Bibcode:1997RvMP...69..995W. doi:10.1103/RevModPhys.69.995.

[23] Earnshaw, A; Greenwood, N (1997). *Chemistry of the Elements* (2nd ed.). Butterworth-Heinemann.

[24] Croswell, K (1996). *Alchemy of the Heavens*. Anchor. ISBN 0-385-47214-5.

[25] Plato (2008) [c. 360 BC]. *Timaeus*. Forgotten Books. p. 45. ISBN 978-1-60620-018-6.

[26] Hillar, M (2004). "The Problem of the Soul in Aristotle's De anima". NASA/WMAP. Retrieved 10 August 2006.

[27] Partington, JR (1937). *A Short History of Chemistry*. New York: Dover Publications. ISBN 0-486-65977-1.

[28] Boyle, R (1661). *The Sceptical Chymist*. London. ISBN 0-922802-90-4.

[29] Lavoisier, AL (1790). *Elements of chemistry translated by Robert Kerr*. Edinburgh. pp. 175–6. ISBN 978-0-415-17914-0.

[30] Transactinide-2. www.kernchemie.de

[31] Carey, GW (1914). *The Chemistry of Human Life*. Los Angeles. ISBN 0-7661-2840-7.

[32] Glanz, J (6 April 2010). "Scientists Discover Heavy New Element". *New York Times.*

[33] Greiner, W. "Recommendations" (PDF). *31st meeting, PAC for Nuclear Physics*. Joint Institute for Nuclear Research.

[34] "IUPAC Announces Start of the Name Approval Process for the Element of Atomic Number 112" (PDF). IUPAC. 20 July 2009. Retrieved 27 August 2009.

[35] "IUPAC (International Union of Pure and Applied Chemistry): Element 112 is Named Copernicium". IUPAC. 20 February 2010.

[36] Oganessian, Yu. Ts.; Utyonkov, V.; Lobanov, Yu.; Abdullin, F.; Polyakov, A.; Sagaidak, R.; Shirokovsky, I.; Tsyganov, Yu. et al. (2006). "Evidence for Dark Matter" (PDF). *Physical Review C* **74** (4): 044602. Bibcode:2006PhRvC..74d4602O. doi:10.1103/PhysRevC.74.044602.

[37] "Two ultra-heavy elements added to the periodic table". 6 June 2011.

7.9   Further reading

- Ball, P (2004). *The Elements: A Very Short Introduction*. Oxford University Press. ISBN 0-19-284099-1.

- Emsley, J (2003). *Nature's Building Blocks: An A-Z Guide to the Elements*. Oxford University Press. ISBN 0-19-850340-7.

- Gray, T (2009). *The Elements: A Visual Exploration of Every Known Atom in the Universe*. Black Dog & Leventhal Publishers Inc. ISBN 1-57912-814-9.

- Scerri, ER (2007). *The Periodic Table, Its Story and Its Significance*. Oxford University Press.

- Strathern, P (2000). *Mendeleyev's Dream: The Quest for the Elements*. Hamish Hamilton Ltd. ISBN 0-241-14065-X.

- Kean, Sam (2011). *The Disappearing Spoon: And Other True Tales of Madness, Love, and the History of the World from the Periodic Table of the Elements*. Back Bay Books.

- Compiled by A. D. McNaught and A. Wilkinson. (1997). Blackwell Scientific Publications, Oxford, ed. *Compendium of Chemical Terminology, 2nd ed. (the "Gold Book")*. doi:10.1351/goldbook. ISBN 0-9678550-9-8.

  XML on-line corrected version: created by M. Nic, J. Jirat, B. Kosata; updates compiled by A. Jenkins.

7.10   External links

- Videos for each element by the University of Nottingham

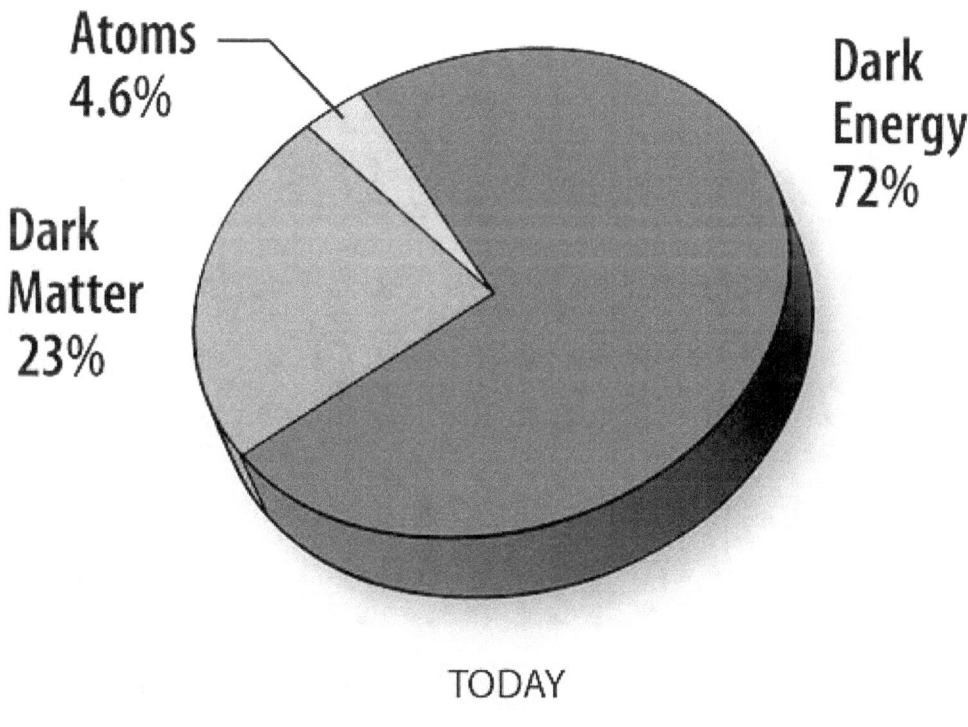

TODAY

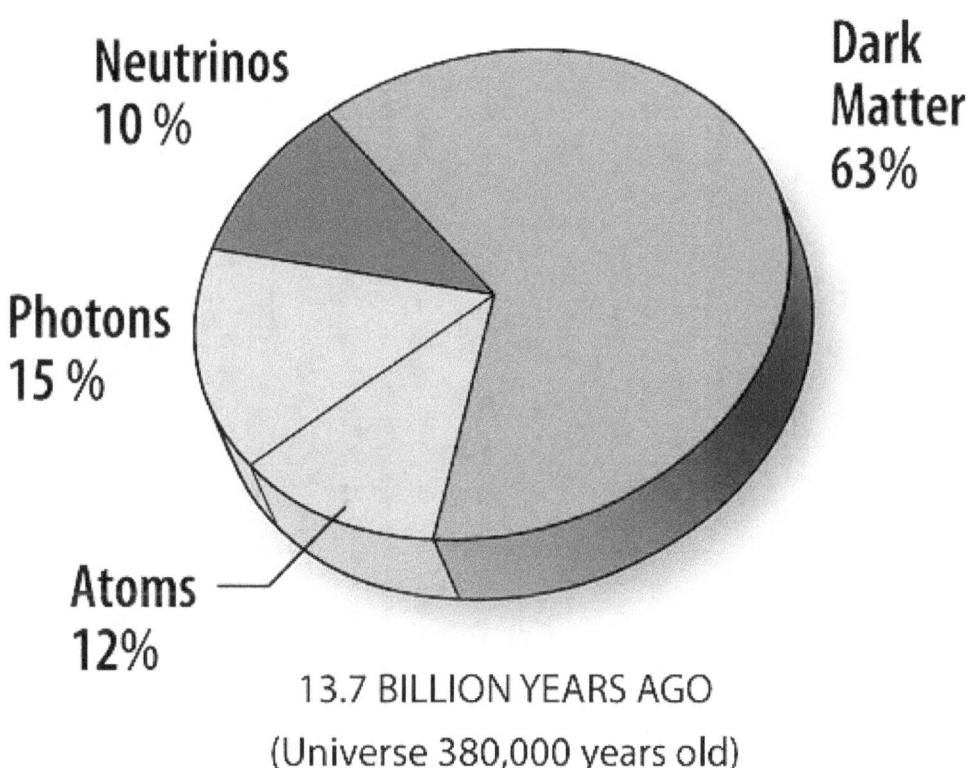

Estimated distribution of dark matter and dark energy in the universe. Only the fraction of the mass and energy in the universe labeled "atoms" is composed of chemical elements.

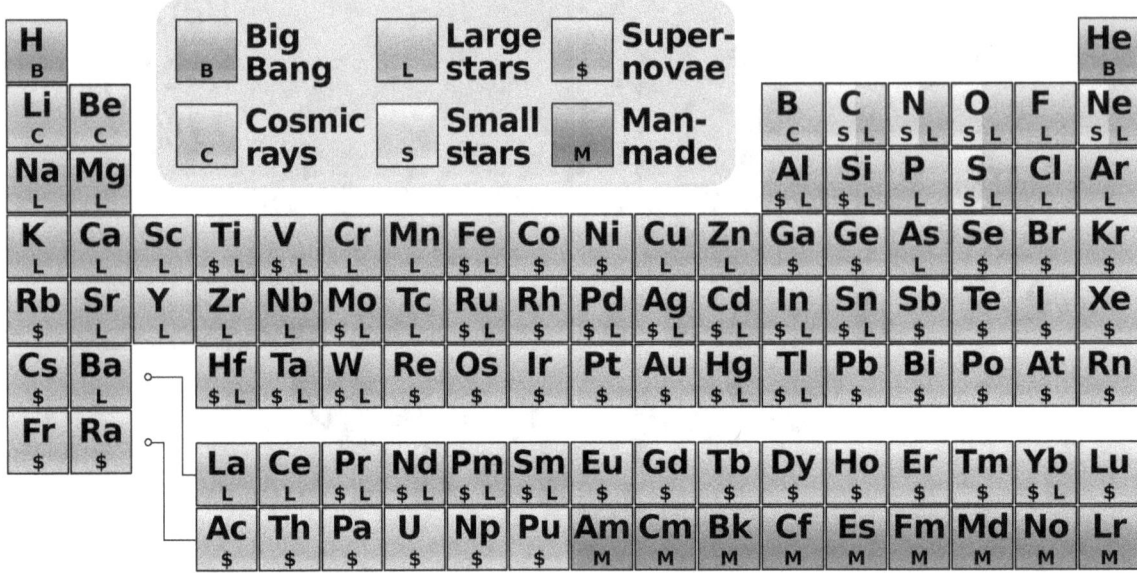

Periodic table showing the cosmogenic origin of each element in the Big Bang, or in large or small stars. Small stars can produce certain elements up to sulfur, by the alpha process. Supernovae are needed to produce "heavy" elements (those beyond iron and nickel) rapidly by neutron buildup, in the r-process. Certain large stars slowly produce other elements heavier than iron, in the s-process; these may then be blown into space in the off-gassing of planetary nebulae

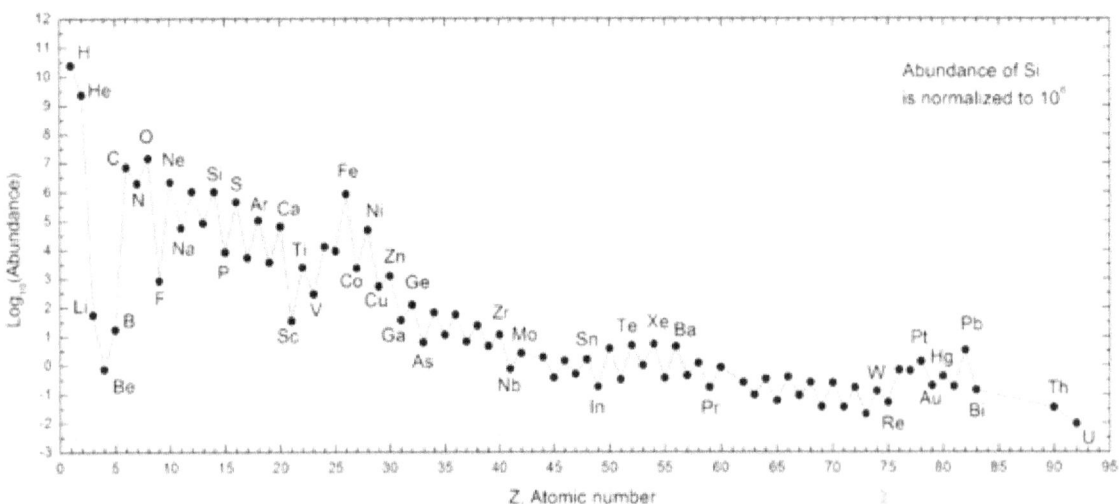

Abundances of the chemical elements in the Solar system. Hydrogen and helium are most common, from the Big Bang. The next three elements (Li, Be, B) are rare because they are poorly synthesized in the Big Bang and also in stars. The two general trends in the remaining stellar-produced elements are: (1) an alternation of abundance in elements as they have even or odd atomic numbers (the Oddo-Harkins rule), and (2) a general decrease in abundance as elements become heavier. Iron is especially common because it represents the minimum energy nuclide that can be made by fusion of helium in supernovae.

ОПЫТЪ СИСТЕМЫ ЭЛЕМЕНТОВЪ,

ОСНОВАННОЙ НА ИХЪ АТОМНОМЪ ВѢСѢ И ХИМИЧЕСКОМЪ СХОДСТВѢ.

			Ti=50	Zr=90	?=180.
			V=51	Nb=94	Ta=182.
			Cr=52	Mo=96	W=186.
			Mn=55	Rh=104,4	Pt=197,1.
			Fe=56	Ru=104,4	Ir=198.
		Ni=Co=59	Pd=106,6	Os=199.
H=1			Cu=63,4	Ag=108	Hg=200.
	Be= 9,4	Mg=24	Zn=65,2	Cd=112	
	B=11	Al=27,3	?=68	Ur=116	Au=197?
	C=12	Si=28	?=70	Sn=118	
	N=14	P=31	As=75	Sb=122	Bi=210?
	O=16	S=32	Se=79,4	Te=128?	
	F=19	Cl=35,5	Br=80	I=127	
Li=7	Na=23	K=39	Rb=85,4	Cs=133	Tl=204.
		Ca=40	Sr=87,6	Ba=137	Pb=207.
		?=45	Ce=92		
		?Er=56	La=94		
		?Yt=60	Di=95		
		?In=75,6	Th=118?		

Д. Менделѣевъ

Mendeleev's 1869 periodic table: An experiment on a system of elements. Based on their atomic weights and chemical similarities.

Dmitri Mendeleev

Henry Moseley

Chapter 8

Chemical nomenclature

A **chemical nomenclature** is a set of rules to generate systematic names for chemical compounds. The nomenclature used most frequently worldwide is the one created and developed by the International Union of Pure and Applied Chemistry (IUPAC).

The IUPAC's rules for naming organic and inorganic compounds are contained in two publications, known as the *Blue Book*[1][2] and the *Red Book*,[3] respectively. A third publication, known as the *Green Book*,[4] describes the recommendations for the use of symbols for physical quantities (in association with the IUPAP), while a fourth, the *Gold Book*,[5] contains the definitions of a large number of technical terms used in chemistry. Similar compendia exist for biochemistry[6] (the *White Book*, in association with the IUBMB), analytical chemistry[7] (the *Orange Book*), macromolecular chemistry[8] (the *Purple Book*) and clinical chemistry[9] (the *Silver Book*). These "color books" are supplemented by shorter recommendations for specific circumstances that are published from time to time in the journal *Pure and Applied Chemistry*.

8.1 Aims of chemical nomenclature

The primary function of chemical nomenclature is to ensure that a spoken or written chemical name leaves no ambiguity concerning which chemical compound the name refers to: each chemical name should refer to a single substance. A less important aim is to ensure that each substance has a single name, although a limited number of alternative names is acceptable in some cases.

Preferably, the name also conveys some information about the structure or chemistry of a compound. CAS numbers form an extreme example of names that do not perform this function: each CAS number refers to a single compound but none contain information about the structure.

The form of nomenclature used depends on the audience to which it is addressed. As such, no single *correct* form exists, but rather there are different forms that are more or less appropriate in different circumstances.

A common name will often suffice to identify a chemical compound in a particular set of circumstances. To be more generally applicable, the name should indicate at least the chemical formula. To be more specific still, the three-dimensional arrangement of the atoms may need to be specified.

In a few specific circumstances (such as the construction of large indices), it becomes necessary to ensure that each compound has a unique name: This requires the addition of extra rules to the standard IUPAC system (the CAS system is the most commonly used in this context), at the expense of having names that are longer and less familiar to most readers. Another system gaining popularity is the International Chemical Identifier (InChI)— which reflects a substance's structure and composition, making it more general than a CAS number.

The IUPAC system is often criticized for the above failures when they become relevant (for example, in differing reactivity of sulfur allotropes, which IUPAC does not distinguish). While IUPAC has a human-readable advantage over CAS numbering, it would be difficult to claim that the IUPAC names for some larger, relevant molecules (such as rapamycin) are human-readable, and so most researchers simply use the informal names.

8.2 Differing aims of chemical nomenclature and lexicography

It is generally understood that the aims of lexicography versus chemical nomenclature vary and are to an extent at odds. Dictionaries of words, whether in traditional print or on the web, collect and report the meanings of words as their uses appear and change over time. For web dictionaries with limited or no formal editorial process, definitions—in this case, definitions of chemical names and terms—can change rapidly without concern for the formal or historical meanings. Chemical nomenclature on the other hand (with IUPAC nomenclature as the best example) is necessarily more restrictive: It aims to standardize communication and practice so that, when a chemical term is used it has a fixed meaning relating to chemical structure, thereby giving insights into chemical properties and derived molecular functions. These differing aims can have profound effects on valid understanding in chemistry, especially with regard to chemical classes that have achieved mass attention. Examples of the impact of these can be seen in considering the examples of:

- resveratrol, a single compound clearly defined by this common name, but that can be confused, popularly, with its cis-isomer,

- omega-3 fatty acids, a reasonably well-defined chemical structure class that is nevertheless broad as a result of its formal definition, and

- polyphenols, a fairly broad structural class with a formal definition, but where mistranslations and general misuse of the term relative to the formal definition has led to serious usage errors, and so ambiguity in the relationship between structure and activity (SAR).

The rapid pace at which meanings can change on the web, in particular for chemical compounds with perceived health benefits, rightly or wrongly ascribed, complicates the matter of maintaining a sound nomenclature (and so access to SAR understanding). A further discussion with specific examples appears in the article on polyphenols, where differing definitions are in use, and there are various, further web definitions and common uses of the word odds with any accepted chemical nomenclature connecting polyphenol structure and bioactivity).

8.3 History

The nomenclature of alchemy is rich in description, but does not effectively meet the aims outlined above. Opinions differ about whether this was deliberate on the part of the early practitioners of alchemy or whether it was a consequence of the particular (and often esoteric) theoretical framework in which they worked.

While both explanations are probably valid to some extent, it is remarkable that the first "modern" system of chemical nomenclature appeared at the same time as the distinction (by Lavoisier) between elements and compounds, in the late eighteenth century.

The French chemist Louis-Bernard Guyton de Morveau published his recommendations[10] in 1782, hoping that his "constant method of denomination" would "help the intelligence and relieve the memory". The system was refined in collaboration with Berthollet, de Fourcroy and Lavoisier,[11] and promoted by the latter in a textbook that would survive long after his death at the guillotine in 1794.[12] The project was also espoused by Jöns Jakob Berzelius,[13][14] who adapted the ideas for the German-speaking world.

The recommendations of Guyton covered only what would be today known as inorganic compounds. With the massive expansion of organic chemistry in the mid-nineteenth century and the greater understanding of the structure of organic compounds, the need for a less *ad hoc* system of nomenclature was felt just as the theoretical tools became available to make this possible. An international conference was convened in Geneva in 1892 by the national chemical societies, from which the first widely accepted proposals for standardization arose.[15]

A commission was set up in 1913 by the Council of the International Association of Chemical Societies, but its work was interrupted by World War I. After the war, the task passed to the newly formed International Union of Pure and Applied Chemistry, which first appointed commissions for organic, inorganic, and biochemical nomenclature in 1921 and continues to do so to this day.

8.4 Types of nomenclature

8.4.1 Organic chemistry

Main article: IUPAC nomenclature of organic chemistry

- Substitutive name

- Functional class name, also known as a radicofunctional name

- Conjunctive name

- Additive name

- Subtractive name

- Multiplicative name

- Fusion name

- Hantzsch–Widman name

- Replacement name

8.4.2 Inorganic chemistry

Main article: IUPAC nomenclature of inorganic chemistry

Compositional nomenclature

Type-I ionic binary compounds For type-I ionic binary compounds, the cation (a metal in most cases) is named first, and the anion (usually a nonmetal) is named second. The cation retains its elemental name (e.g., *iron* or *zinc*), but the suffix of the nonmetal changes to *-ide*. For example, the compound LiBr is made of Li^+ cations and Br^- anions; thus, it's called lithium bromide. The compound BaO, which is composed of Ba^{2+} cations and O^{2-} anions, is referred to as barium oxide.

The oxidation state of each element is unambiguous. When these ions combine into a type-I binary compound, their equal-but-opposite charges are neutralized, so the compound's net charge is zero.

Type-II ionic binary compounds Type-II ionic binary compounds are those in which the cation does not have just one oxidation state. This is common among transition metals. To name these compounds, one must determine the charge of the cation and then write out the name as would be done with Type I Ionic Compounds, except that a Roman numeral (indicating the charge of the cation) is written in parentheses next to the cation name (this is sometimes referred to as Stock nomenclature). For example, take the compound $FeCl_3$. The cation, iron, can occur as Fe^{2+} and Fe^{3+}. In order for the compound to have a net charge of zero, the cation must be Fe^{3+} so that the three Cl^- anions can be balanced out (3+ and 3− balance to 0). Thus, this compound is called iron(III) chloride. Another example could be the compound PbS_2. Because the S^{2-} anion has a subscript of 2 in the formula (giving a 4− charge), the compound must be balanced with a 4+ charge on the Pb cation (lead is a transition metal, and can form cations with a 4+ or a 2+ charge). Thus, the compound is made of one Pb^{4+} cation to every two S^{2-} anions, the compound is balanced, and its name is written as lead(IV) sulfide. An older system – relying on Latin names for the elements – is also sometimes used to name Type II Ionic Binary Compounds. In this system, the metal (instead of a Roman numeral next to it) has an "-ic" or "-ous" suffix added to it to indicate its oxidation state ("-ous" for lower, "-ic" for higher). For example, the compound FeO contains the Fe^{2+} cation (which balances out with the O^{2-} anion). Since this oxidation state is lower than the other possibility (Fe^{3+}), this compound is sometimes called ferrous oxide. For the compound, SnO_2, the tin ion is Sn^{4+} (balancing out the 4– charge on the two O^{2-} anions), and because this is a higher oxidation state than the alternative (Sn^{2+}), this compound is called stannic oxide.

Some ionic compounds contain polyatomic ions, which are charged entities containing two or more covalently bonded types of atoms. It is important to know the names of common polyatomic ions; these include:

- ammonium (NH_4^+)

- nitrite (NO_2^-)

- nitrate (NO_3^-)

- sulfite (SO_3^{2-})

- sulfate (SO_4^{2-})

- hydrogen sulfate (bisulfate) (HSO_4^-)

- hydroxide (OH^-)

- cyanide (CN^-)

- phosphate (PO_4^{3-})

- hydrogen phosphate (HPO_4^{2-})

- dihydrogen phosphate ($H_2PO_4^-$)

- carbonate (CO_3^{2-})

- hydrogen carbonate (bicarbonate) (HCO_3^-)

- hypochlorite (ClO^-)

- chlorite (ClO_2^-)

- chlorate (ClO_3^-)

- perchlorate (ClO_4^-)

- acetate ($C_2H_3O_2^-$), permanganate (MnO_4^-)

- dichromate ($Cr_2O_7^{2-}$)

- chromate (CrO_4^{2-})

- peroxide (O_2^{2-})

The formula Na_2SO_3 denotes that the cation is sodium, or Na^+, and that the anion is the sulfite ion (SO_3^{2-}). Therefore, this compound is named sodium sulfite. If the given formula is $Ca(OH)_2$, it can be seen that OH^- is the hydroxide ion. Since the charge on the calcium ion is 2+, it makes sense there must be two OH^- ions to balance the charge. Therefore, the name of the compound is calcium hydroxide. If one is asked to write the formula for copper(I) chromate, the Roman numeral indicates that copper ion is Cu^+ and one can identify that the compound contains the chromate ion (CrO_4^{2-}). Two of the 1+ copper ions are needed to balance the charge of one 2– chromate ion, so the formula is Cu_2CrO_4.

Type-III binary compounds Type-III binary compounds are covalently bonded. Covalent bonding occurs between nonmetal elements. Covalently-bonded compounds are also known as *molecules*. In the compound, the first element is named first and with its full elemental name. The second element is named as if it were an anion (root name of the element + *-ide* suffix). Then, prefixes are used to indicate the numbers of each atom present: these prefixes are *mono-* (one), *di-* (two), *tri-* (three), *tetra-* (four), *penta-* (five), *hexa-* (six), *hepta-* (seven), *octa-* (eight), *nona-* (nine), and *deca-* (ten). The prefix *mono-* is never used with the first element. Thus, NCl_3 is called nitrogen trichloride, P_2O_5 is called diphosphorus pentoxide (the *a* of the *penta-* prefix is dropped before the vowel for easier pronunciation), and BF_3 is called boron trifluoride.

Carbon dioxide is written CO_2; sulfur tetrafluoride is written SF_4. A few compounds, however, have common names that prevail. H_2O, for example, is usually called *water* rather than *dihydrogen monoxide*, and NH_3 is preferentially called *ammonia* rather than *hydrogen nitride*.

Substitutive nomenclature

This naming method generally follows established IUPAC organic nomenclature. Hydrides of the main group elements (groups 13–17) are given -ane base name, e.g. borane (BH_3), oxidane (H_2O), phosphane (PH_3) (Although the name *phosphine* is also in common use, it is not recommended by IUPAC). The compound PCl_3 would thus be named substitutively as trichlorophosphane (with chlorine "substituting"). However, not all such names (or stems) are derived from the element name. For example, NH_3 is called "azane" (rather than a word such as *nitro-ane* which is rather difficult to pronounce in English).

Additive nomenclature

This naming method has been developed principally for coordination compounds although it can be more widely applied. An example of its application is $[CoCl(NH_3)_5]Cl_2$ pentaamminechloridocobalt(III) chloride.

Ligands, too, have a special naming convention. Whereas *chloride* becomes the prefix *chloro-* in substitutive naming, in a ligand it becomes *chlorido-*.

8.5 See also

- IUPAC nomenclature of inorganic chemistry 2005
- IUPAC nomenclature of organic chemistry
- Preferred IUPAC name
- IUPAC numerical multiplier
- IUPAC nomenclature for organic transformations
- International Chemical Identifier
- List of chemical compounds with unusual names

8.6 References

[1] *Nomenclature of Organic Chemistry* (3rd ed.), London: Butterworths, 1971 [1958 (A: Hydrocarbons, and B: Fundamental Heterocyclic Systems), 1965 (C: Characteristic Groups)], ISBN 0-408-70144-7.

[2] Rigaudy, J.; Klesney, S. P., eds. (1979). *Nomenclature of Organic Chemistry*. IUPAC/Pergamon Press. ISBN 0-08022-3699.. Panico R, Powell WH, Richer JC, eds. (1993). *A Guide to IUPAC Nomenclature of Organic Compounds*. IUPAC/Blackwell Science. ISBN 0-632-03488-2.. IUPAC, Chemical Nomenclature and Structure Representation Division (27 October 2004). *Nomenclature of Organic Chemistry (Provisional Recommendations)*. IUPAC..

[3] International Union of Pure and Applied Chemistry (2005). *Nomenclature of Inorganic Chemistry* (IUPAC Recommendations 2005). Cambridge (UK): RSC–IUPAC. ISBN 0-85404-438-8. Electronic version..

[4] International Union of Pure and Applied Chemistry (1993). *Quantities, Units and Symbols in Physical Chemistry*, 2nd edition, Oxford: Blackwell Science. ISBN 0-632-03583-8. Electronic version..

[5] *Compendium of Chemical Terminology, IMPACT Recommendations (2nd Ed.)*, Oxford:Blackwell Scientific Publications. (1997)

[6] *Biochemical Nomenclature and Related Documents*, London:Portland Press, 1992.

[7] International Union of Pure and Applied Chemistry (1998). *Compendium of Analytical Nomenclature* (definitive rules 1997, 3rd. ed.). Oxford: Blackwell Science. ISBN 0-86542-6155. .

[8] *Compendium of Macromolecular Nomenclature*, Oxford:Blackwell Scientific Publications, 1991.

[9] *Compendium of Terminology and Nomenclature of Properties in Clinical Laboratory Sciences*, IMPACT Recommendations 1995, Oxford: Blackwell Science, ISBN 0-86542-612-0.

[10] Guyton de Morveau, L. B. (1782), *J. Phys.* **19**: 310 Missing or empty |title= (help).

[11] Guyton de Morveau, L. B.; Lavoisier, A. L.; Berthollet, C. L.; Fourcroy, A. F. de (1787), *Méthode de Nomenclature Chimique*, Paris: Cuchet.

[12] Lavoisier, A. L. (1801), *Traité Élémentaire de Chimie* (3e ed.), Paris: Deterville.

[13] Berzelius, J. J. (1811), *J. Phys.* **73**: 248 Missing or empty |title= (help).

[14]Wisniak, Jaime (2000), "Jöns Jacob Berzelius A Guide to the Perplexed Chemist",*Chem. Educator***5**(6): 343–50,doi:10.1007/.

[15] "Congrès de nomenclature chimique, Genève 1892", *Bull. Soc. Chim. Paris, Ser. 3* **7**, 1892: xiii–xxiv.

8.7 External links

- Interactive IUPAC Compendium of Chemical Terminology (interactive "Gold Book")

- IUPAC Nomenclature Books Series (list of all IUPAC nomenclature books, and means of accessing them)

- IUPAC Compendium of Chemical Terminology ("*Gold Book*")

- Quantities, Units and Symbols in Physical Chemistry ("*Green Book*")

- IUPAC Nomenclature of Organic Chemistry ("*Blue Book*")

- Nomenclature of Inorganic Chemistry IUPAC Recommendations 2005 ("*Red Book*")

- IUPAC Recommendations on Organic & Biochemical Nomenclature, Symbols, Terminology, etc.(includes IUBM Recommendations for biochemistry)

- chemicalize.org A free web site/service that extracts IUPAC names from web pages and annotates a 'chemicalized' version with structure images. Structures from annotated pages can also be searched.

- ChemAxon Name <> Structure – IUPAC (& traditional) name to structure and structure to IUPAC name software. As used at chemicalize.org

- ACD/Name – Generates IUPAC, INDEX (CAS), InChi, Smiles, etc. for drawn structures in 10 languages and translates names to structures. Also available as batch tool and for Pipeline Pilot. Part of I-Lab 2.0

CHYMICAL

NOMENCLATURE.

A MEMOIR.

ON THE NECESSITY OF REFORMING AND BRINGING TO PERFECTION THE NOMENCLATURE OF CHYMISTRY; READ TO THE PUBLIC ASSEMBLY OF THE ROYAL ACADEMY OF SCIENCES IN PARIS, ON THE 18th OF APRIL, 1787.

By Mr. LAVOISIER.

THE work which we lay before the Academy has been undertaken in common by Mr. de Morveau, Mr. Bertholet, Mr. de Fourcroy, and by me: it is the refult of a great number of confultations, in which we have been affifted by the learning and advice of fome geometricians of the Academy, and of feveral chymifts.

A long time before the modern difcoveries had given a new appearance to the fcience in general, chymifts perceived the neceffity of giving the no-

B   menclature

Chapter 9

Chemical formula

*Structural formula for butane. This is **not** a chemical formula. Examples of chemical formulas for butane are the empirical formula C_2H_5, the **molecular formula** C_4H_{10} and the condensed (or semi-structural) formula $CH_3CH_2CH_2CH_3$.*

Not to be confused with the 2-D graphical method of showing atomic spacial relationships called a structural formula.

A **chemical formula** is a way of expressing information about the proportions of atoms that constitute a particular chemical compound, using a single line of chemical element symbols, numbers, and sometimes also other symbols, such as parentheses, dashes, brackets, commas and *plus* (+) and *minus* (−) signs. These are limited to a single typographic line of symbols, which may include subscripts and superscripts. A chemical formula is not a chemical name, and it contains no words. Although a chemical formula may imply certain simple chemical structures, it is not the same as a full chemical structural formula. Chemical formulas can fully specify the structure of only the simplest of molecules and chemical substances, and are generally more limited in power than are chemical names and structural formulas.

The simplest types of chemical formulas are called *empirical formulas*, which use only letters and numbers indicating atomic proportional ratios (the numerical proportions of atoms of one type to those of other types). *Molecular formulas* indicate the simple numbers of each type of atom in a molecule of a molecular substance, and are thus sometimes the same as empirical formulas (for molecules that only have one atom of a particular type), and at other times require larger numbers than do empirical formulas. An example of the difference is the empirical formula for glucose, which is CH_2O, while its molecular formula requires all numbers to be increased by a factor of six, giving $C_6H_{12}O_6$.

Sometimes a chemical formula is complicated by being written as a condensed formula (or condensed molecular formula, occasionally called a "semi-structural formula"), which conveys additional information about the particular ways in which the atoms are chemically bonded together, either in covalent bonds, ionic bonds, or various combinations of these types. This is possible if the relevant bonding is easy to show in one dimension. An example is the condensed molecular/chemical formula for ethanol, which is CH_3-CH_2-OH or CH_3CH_2OH. However, even a condensed chemical formula is necessarily limited in its ability to show complex bonding relationships between atoms, especially atoms that have bonds to four or more different substituents.

Since a chemical formula must be expressed as a single line of chemical element symbols, it often cannot be as informative as a true structural formula, which is a graphical representation of the spacial relationship between atoms in chemical compounds (see for example the figure for butane structural and chemical formulas, at right). For reasons of structural complexity, there is no condensed chemical formula (or semi-structural formula) that specifies glucose (and there exist many different molecules, for example fructose and mannose, have the same molecular formula $C_6H_{12}O_6$ as glucose). Linear equivalent chemical *names* exist that can and do specify any complex structural formula (see chemical nomenclature), but such names must use many terms (words), rather than the simple element symbols, numbers, and simple typographical symbols that define a chemical formula.

Chemical formulas may be used in chemical equations to describe chemical reactions and other chemical transformations, such as the dissolving of ionic compounds into solution. While, as noted, chemical formulas do not have the full power of structural formulas to show chemical relationships between atoms, they are sufficient to keep track of numbers of atoms and numbers of electrical charges in chemical reactions, thus balancing chemical equations so that these equations can be used in chemical problems involving conservation of atoms, and conservation of electric charge.

9.1 Overview

A chemical formula identifies each constituent element by its chemical symbol and indicates the proportionate number of atoms of each element. In empirical formulas, these proportions begin with a key element and then assign numbers of atoms of the other elements in the compound, as ratios to the key element. For molecular compounds, these ratio numbers can all be expressed as whole numbers. For example, the empirical formula of ethanol may be written C_2H_6O because the molecules of ethanol all contain two carbon atoms, six hydrogen atoms, and one oxygen atom. Some types of ionic compounds, however, cannot be written with entirely whole-number empirical formulas. An example is boron carbide, whose formula of CB_n is a variable non-whole number ratio with n ranging from over 4 to more than 6.5.

When the chemical compound of the formula consists of simple molecules, chemical formulas often employ ways to suggest the structure of the molecule. These types of formulas are variously known as **molecular formulas** and **condensed formulas**. A molecular formula enumerates the number of atoms to reflect those in the molecule, so that the molecular formula for glucose is $C_6H_{12}O_6$ rather than the glucose empirical formula, which is CH_2O. However, except for very simple substances, molecular chemical formulas lack needed structural information, and are ambiguous.

For simple molecules, a condensed (or semi-structural) formula is a type of chemical formula that may fully imply a correct structural formula. For example, ethanol may be represented by the condensed chemical formula CH_3CH_2OH, and dimethyl ether by the condensed formula CH_3OCH_3. These two molecules have the same empirical and molecular formulas (C_2H_6O), but may be differentiated by the condensed formulas shown, which are sufficient to represent the full structure of these simple organic compounds.

Condensed chemical formulas may also be used to represent ionic compounds that do not exist as discrete molecules, but nonetheless do contain covalently bound clusters within them. These polyatomic ions are groups of atoms that are covalently bound together and have an overall ionic charge, such as the sulfate [SO
4]2−
ion. Each polyatomic ion in a compound is written individually in order to illustrate the separate groupings. For example, the compound dichlorine hexoxide has an empirical formula ClO
3, and molecular formula Cl
2O
6, but in liquid or solid forms, this compound is more correctly shown by an ionic condensed formula [ClO
2]+

[ClO
4]−
, which illustrates that this compound consists of [ClO
2]+
ions and [ClO
4]−
ions. In such cases, the condensed formula only need be complex enough to show at least one of each ionic species.

Chemical formulas must be differentiated from the far more complex chemical systematic names that are used in various systems of chemical nomenclature. For example, one systematic name for glucose is (2R,3S,4R,5R)−2,3,4,5,6-pentahydroxyhexanal. This name, and the rules behind it, fully specify glucose's structural formula, but the name is not a chemical formula, as it uses many extra terms and words that chemical formulas do not permit. Such chemical names may be able to represent full structural formulas without graphs, but in order to do so, they require word terms that are not part of chemical formulas.

9.2   Simple empirical formulas

Main article: Empirical formula

In chemistry, the empirical formula of a chemical is a simple expression of the relative number of each type of atom or ratio of the elements in the compound. Empirical formulas are the standard for ionic compounds, such as CaCl
2, and for macromolecules, such as SiO
2. An empirical formula makes no reference to isomerism, structure, or absolute number of atoms. The term **empirical** refers to the process of elemental analysis, a technique of analytical chemistry used to determine the relative percent composition of a pure chemical substance by element.

For example hexane has a molecular formula of C
6H
14, or structurally CH
3CH
2CH
2CH
2CH
2CH
3, implying that it has a chain structure of 6 carbon atoms, and 14 hydrogen atoms. However, the empirical formula for hexane is C
3H
7. Likewise the empirical formula for hydrogen peroxide, H
2O
2, is simply HO expressing the 1:1 ratio of component elements. Formaldehyde and acetic acid have the same empirical formula, CH
2O. This is the actual chemical formula for formaldehyde, but acetic acid has double the number of atoms.

9.3   Condensed formulas in organic chemistry implying molecular geometry and structural formulas

The connectivity of a molecule often has a strong influence on its physical and chemical properties and behavior. Two molecules composed of the same numbers of the same types of atoms (i.e. a pair of isomers) might have completely different chemical and/or physical properties if the atoms are connected differently or in different positions. In such cases, a structural formula is useful, as it illustrates which atoms are bonded to which other ones. From the connectivity, it is often possible to deduce the approximate shape of the molecule.

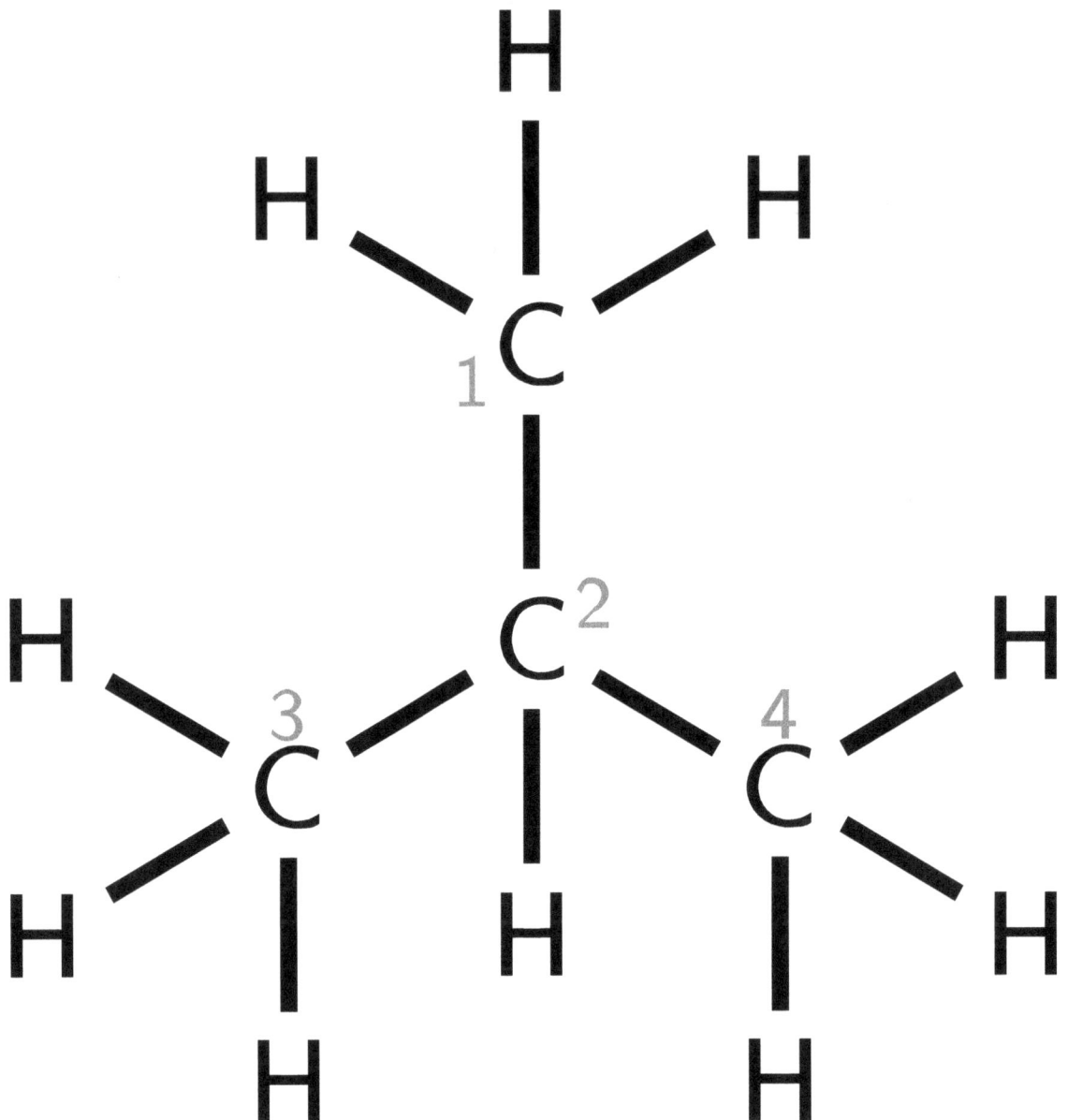

Isobutane structural formula
Molecular formula: C_4H_{10}
Condensed or semi-structural chemical formula: $(CH_3)_3CH$

A condensed chemical formula may represent the types and spatial arrangement of bonds in a simple chemical substance, though it does not necessarily specify isomers or complex structures. For example ethane consists of two carbon atoms single-bonded to each other, with each carbon atom having three hydrogen atoms bonded to it. Its chemical formula can be rendered as CH_3CH_3. In ethylene there is a double bond between the carbon atoms (and thus each carbon only has two hydrogens), therefore the chemical formula may be written: CH_2CH_2, and the fact that there is a double bond between the carbons is implicit because carbon has a valence of four. However, a more explicit method is to write $H_2C=CH_2$ or less commonly $H_2C::CH_2$. The two lines (or two pairs of dots) indicate that a double bond connects the atoms on either side of them.

A triple bond may be expressed with three lines or pairs of dots, and if there may be ambiguity, a single line or pair of dots may be used to indicate a single bond.

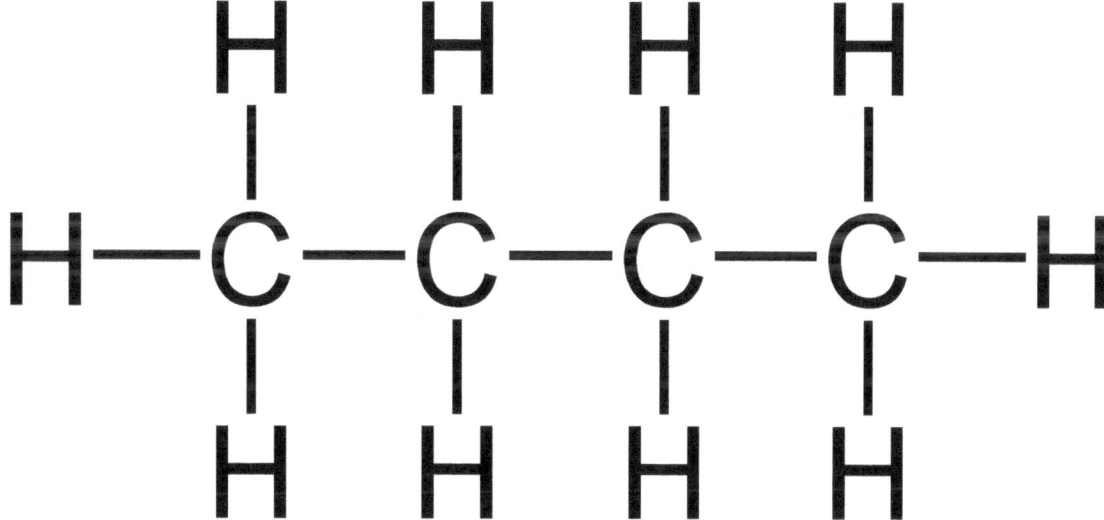

Butane structural formula
Molecular formula: C_4H_{10}
Condensed or semi-structural formula: $CH_3CH_2CH_2CH_3$

Molecules with multiple functional groups that are the same may be expressed by enclosing the repeated group in round brackets. For example isobutane may be written $(CH_3)_3CH$. This condensed structural formula implies a different connectivity from other molecules that can be formed using the same atoms in the same proportions (isomers). The formula $(CH_3)_3CH$ implies a central carbon atom attached to one hydrogen atom and three CH_3 groups. The same number of atoms of each element (10 hydrogens and 4 carbons, or C_4H_{10}) may be used to make a straight chain molecule, butane: $CH_3CH_2CH_2CH_3$.

9.4   Chemical names in answer to limitations of chemical formulas

Main article: chemical nomenclature

The alkene called **but-2-ene** has two isomers, which the chemical formula $CH_3CH=CHCH_3$ does not identify. The relative position of the two methyl groups must be indicated by additional notation denoting whether the methyl groups are on the same side of the double bond (*cis* or *Z*) or on the opposite sides from each other (*trans* or *E*). Such extra symbols violate the rules for chemical formulas, and begin to enter the territory of more complex naming systems.

As noted above, in order to represent the full structural formulas of many complex organic and inorganic compounds, chemical nomenclature may be needed which goes well beyond the available resources used above in simple condensed formulas. See IUPAC nomenclature of organic chemistry and IUPAC nomenclature of inorganic chemistry 2005 for examples. In addition, linear naming systems such as International Chemical Identifier (InChI) allow a computer to construct a structural formula, and simplified molecular-input line-entry system (SMILES) allows a more human-readable ASCII input. However, all these nomenclature systems go beyond the standards of chemical formulas, and technically are chemical naming systems, not formula systems.

9.5   Polymers in condensed formulas

For polymers in condensed chemical formulas, parentheses are placed around the repeating unit. For example, a hydrocarbon molecule that is described as $CH_3(CH_2)_{50}CH_3$, is a molecule with fifty repeating units. If the number of repeating units is unknown or variable, the letter *n* may be used to indicate this formula: $CH_3(CH_2)nCH_3$.

9.6 Ions in condensed formulas

For ions, the charge on a particular atom may be denoted with a right-hand superscript. For example Na^+, or Cu^{2+}. The total charge on a charged molecule or a polyatomic ion may also be shown in this way. For example: H_3O^+ or SO_4^{2-}.

For more complex ions, brackets [ ] are often used to enclose the ionic formula, as in $[B_{12}H_{12}]^{2-}$, which is found in compounds such as $Cs_2[B_{12}H_{12}]$. Parentheses ( ) can be nested inside brackets to indicate a repeating unit, as in $[Co(NH_3)_6]^{3+}$. Here $(NH_3)_6$ indicates that the ion contains six NH_3 groups, and [ ] encloses the entire formula of the ion with charge +3.

9.7 Isotopes

Although isotopes are more relevant to nuclear chemistry or stable isotope chemistry than to conventional chemistry, different isotopes may be indicated with a prefixed superscript in a chemical formula. For example, the phosphate ion containing radioactive phosphorus-32 is $^{32}PO_4^{3-}$. Also a study involving stable isotope ratios might include the molecule $^{18}O^{16}O$.

A left-hand subscript is sometimes used redundantly to indicate the atomic number. For example, $_8O_2$ for dioxygen, and $^{16}_8O_2$ for the most abundant isotopic species of dioxygen. This is convenient when writing equations for nuclear reactions, in order to show the balance of charge more clearly.

9.8 Trapped atoms

Main article: Endohedral fullerene

The @ symbol (at sign) indicates an atom or molecule trapped inside a cage but not chemically bound to it. For example, a buckminsterfullerene (C_{60}) with an atom (M) would simply be represented as MC_{60} regardless of whether M was inside the fullerene without chemical bonding or outside, bound to one of the carbon atoms. Using the @ symbol, this would be denoted $M@C_{60}$ if M was inside the carbon network. A non-fullerene example is $[As@Ni_{12}As_{20}]^{3-}$, an ion in which one As atom is trapped in a cage formed by the other 32 atoms.

This notation was proposed in 1991[1] with the discovery of fullerene cages (endohedral fullerenes), which can trap atoms such as La to form, for example, $La@C_{60}$ or $La@C_{82}$. The choice of the symbol has been explained by the authors as being concise, readily printed and transmitted electronically (the at sign is included in ASCII, which most modern character encoding schemes are based on), and the visual aspects suggesting the structure of an endohedral fullerene.

9.9 Non-stoichiometric chemical formulas

Main article: Non-stoichiometric compound

Chemical formulas most often use integers for each element. However, there is a class of compounds, called non-stoichiometric compounds, that cannot be represented by small integers. Such a formula might be written using decimal fractions, as in $Fe_{0.95}O$, or it might include a variable part represented by a letter, as in $Fe_{1-x}O$, where x is normally much less than 1.

Traditional formula: MC_{60}
The "@" notation: $M@C_{60}$

9.10   General forms for organic compounds

A chemical formula used for a series of compounds that differ from each other by a constant unit is called **general formula**. Such a series is called the homologous series, while its members are called homologs.
For example alcohols may be represented by: $CnH_{(2n+1)}OH$ ($n \geq 1$)

9.11   Hill System

Main article: Hill system

The **Hill system** is a system of writing chemical formulas such that the number of carbon atoms in a molecule is indicated first, the number of hydrogen atoms next, and then the number of all other chemical elements subsequently, in alphabetical order. When the formula contains no carbon, all the elements, including hydrogen, are listed alphabetically. This deterministic system enables straightforward sorting and searching of compounds.

9.12 See also

- Dictionary of chemical formulas

- Element symbol

- Nuclear notation

- Periodic table

- IUPAC nomenclature of inorganic chemistry

9.13 References

[1] Chai, Yan; Guo, Ting; Jin, Changming; Haufler, Robert E.; Chibante, L. P. Felipe; Fure, Jan; Wang, Lihong; Alford, J. Michael; Smalley, Richard E. (1991). "Fullerenes wlth Metals Inside". *Journal of Physical Chemistry* **95** (20): 7564–7568. doi:10.1021/j100173a002.

- Ralph S. Petrucci, William S. Harwood, F. Geoffrey Herring (2002). "3". *General Chemistry: Principles and Modern Applications* (8th ed.). Prentice-Hall. ISBN 0-13-198825-5. OCLC 46872308.

Chapter 10

Structural formula

The **structural formula** of a chemical compound is a graphic representation of the molecular structure, showing how the atoms are arranged. The chemical bonding within the molecule is also shown, either explicitly or implicitly. Unlike **chemical** formulas, which have a limited number of symbols and are capable of only limited descriptive power, **structural** formulas provide a complete geometric representation of the molecular structure. For example, many chemical compounds exist in different isomeric forms, which have different enantiomeric structures but the same chemical formula. A structural formula is able to indicate arrangements of atoms in three dimensional space in a way that a chemical formula may not be able to do.

Several systematic chemical **naming** formats, as in chemical databases, may be are used that are equivalent to, and as powerful as, geometric structures. These chemical nomenclature systems include SMILES, InChI and CML. These systematic chemical names can be converted to structural formulas and vice versa, but chemists nearly always describe a chemical reaction or synthesis using structural formulas rather than chemical names, because the structural formulas allow the chemist to visualize the molecules and the structural changes that occur in them during chemical reactions.

10.1   Lewis structures

Main article: Lewis structure

Lewis structures (or "Lewis dot structures") are flat graphical formulas that show atom connectivity and lone pair or unpaired electrons, but not three-dimensional structure. This notation is mostly used for small molecules. Each line represents the two electrons of a single bond. Two or three parallel lines between pairs of atoms represent double or triple bonds, respectively. Alternatively, pairs of dots may be used to represent bonding pairs. In addition, all non-bonded electrons (paired or unpaired) and any formal charges on atoms are indicated.

- The Lewis structure of water

10.2   Condensed formulas

In early organic-chemistry publications, where use of graphics was strongly limited, a typographic system arose to describe organic structures in a line of text. Although this system tends to be problematic in application to cyclic compounds, it remains a convenient way to represent simple structures:

CH_3CH_2OH (ethanol)

Parentheses are used to indicate multiple identical groups, indicating attachment to the nearest non-hydrogen atom on the left when appearing within a formula, or to the atom on the right when appearing at the start of a formula:

R = 5'-deoxyadenosyl, Me, OH, CN

Skeletal structural formula of Vitamin B12. Many organic molecules are too complicated to be specified with a chemical formula (molecular formula).

$(CH_3)_2CHOH$ or $CH(CH_3)_2OH$ (2-propanol)

In all cases, all atoms are shown, including hydrogen atoms.

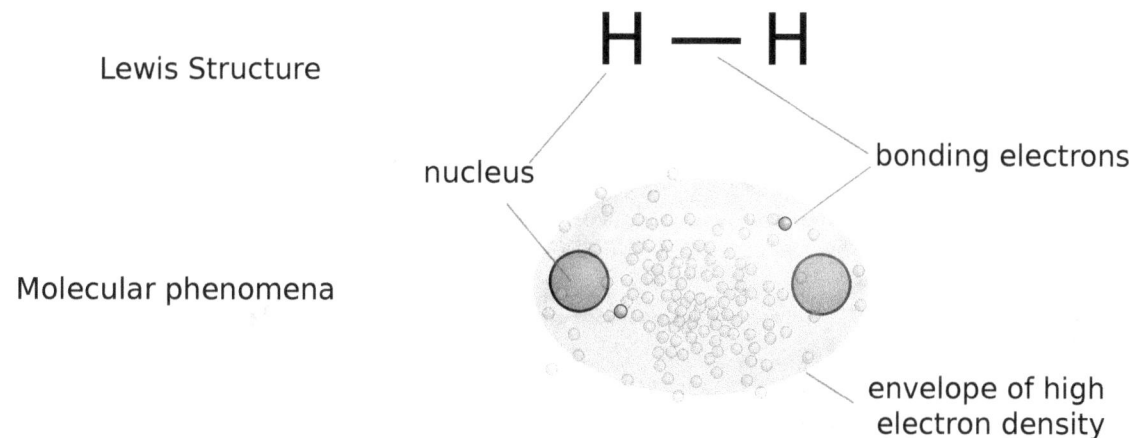

Lewis Structure

Molecular phenomena

Representation of molecules by molecular formula

10.3   Skeletal formulas

Main article: Skeletal formula

Skeletal formulas are the standard notation for more complex organic molecules. First used by the organic chemist Friedrich August Kekulé von Stradonitz the carbon atoms in this type of diagram are implied to be located at the vertices (corners) and termini of line segments rather than being indicated with the atomic symbol C. Hydrogen atoms attached to carbon atoms are not indicated: each carbon atom is understood to be associated with enough hydrogen atoms to give the carbon atom four bonds. The presence of a positive or negative charge at a carbon atom takes the place of one of the implied hydrogen atoms. Hydrogen atoms attached to atoms other than carbon must be written explicitly.

- Skeletal formula of isobutanol

10.4   Indication of stereochemistry

Several methods exist to picture the three-dimensional arrangement of atoms in a molecule (stereochemistry).

10.4.1   Stereochemistry in skeletal formulas

Chirality in skeletal formulas is indicated by the Natta projection method. Solid or dashed wedged bonds represent bonds pointing above-the-plane or below-the-plane of the paper, respectively.

10.4.2   Unspecified stereochemistry

Wavy single bonds represent unknown or unspecified stereochemistry or a mixture of isomers. For example the diagram to the left shows the fructose molecule with a wavy bond to the $HOCH_2$- group at the left. In this case the two possible ring structures are in chemical equilibrium with each other and also with the open-chain structure. The ring continually opens and closes, sometimes closing with one stereochemistry and sometimes with the other.

Skeletal formulae can depict *cis* and *trans* isomers of alkenes. Wavy single bonds are the standard way to represent unknown or unspecified stereochemistry or a mixture of isomers (as with tetrahedeal stereocenters). A crossed double-bond has been used sometimes, but is no longer considered an acceptable style for general use. [1][1]

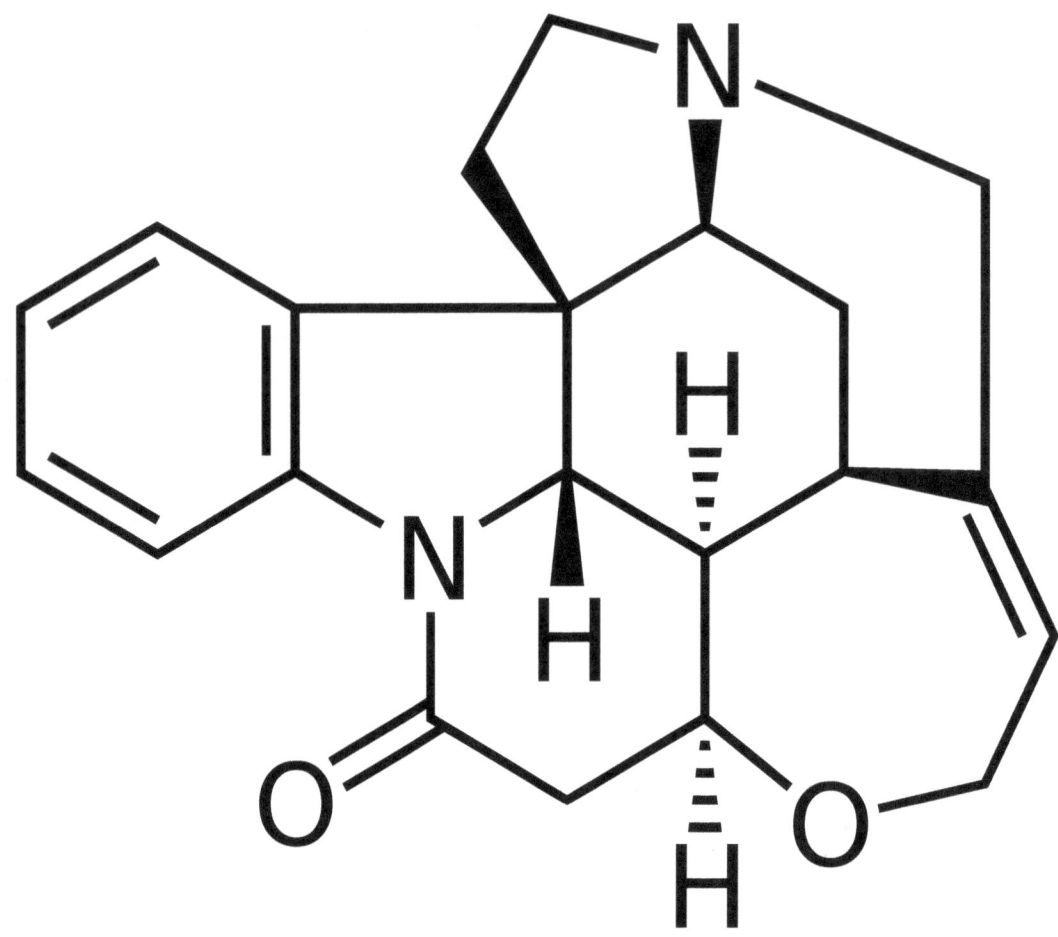

Skeletal formula of strychnine. A solid wedged bond seen for example at the nitrogen (N) at top indicates a bond pointing above-the-plane, while a dashed wedged bond seen for example at the hydrogen (H) at bottom indicates a below-the-plane bond.

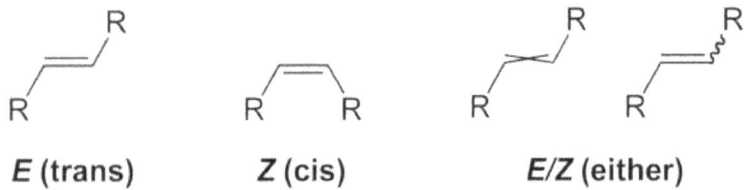

E (trans)   *Z* (cis)   *E/Z* (either)

10.5 Perspective drawings

10.5.1 Newman projection and sawhorse projection

The Newman projection and the sawhorse projection are used to depict specific conformers or to distinguish vicinal stereochemistry. In both cases, two specific carbon atoms and their connecting bond are the center of attention. The only difference is a slightly different perspective: the Newman projection looking straight down the bond of interest, the sawhorse projection looking at the same bond but from a somewhat oblique vantage point. In the Newman projection, a

Fructose, with a bond at the hydroxyl (OH) group upper left of image with unknown or unspecified stereochemistry.

circle is used to represent a plane perpendicular to the bond, distinguishing the substituents on the front carbon from the substituents on the back carbon. In the sawhorse projection, the front carbon is usually on the left and is always slightly lower:

- Newman projection of butane

- sawhorse projection of butane

10.5.2   Cyclohexane conformations

Certain conformations of cyclohexane and other small-ring compounds can be shown using a standard convention. For example, the standard chair conformation of cyclohexane involves a perspective view from slightly above the average plane of the carbon atoms and indicates clearly which groups are axial and which are equatorial. Bonds in front may or may not be highlighted with stronger lines or wedges.

- Chair conformation of beta-D-Glucose

10.5.3   Haworth projection

The Haworth projection is used for cyclic sugars. Axial and equatorial positions are not distinguished; instead, substituents are positioned directly above or below the ring atom to which they are connected. Hydrogen substituents are typically omitted.

- Haworth projection of beta-D-Glucose

10.5.4   Fischer projection

The Fischer projection is mostly used for linear monosaccharides. At any given carbon center, vertical bond lines are equivalent to stereochemical hashed markings, directed away from the observer, while horizontal lines are equivalent to wedges, pointing toward the observer. The projection is totally unrealistic, as a saccharide would never adopt this multiply eclipsed conformation. Nonetheless, the Fischer projection is a simple way of depicting multiple sequential stereocenters that does not require or imply any knowledge of actual conformation:

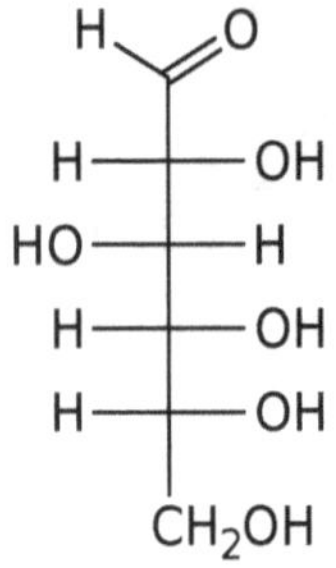

```
      H    O
       \ //
        C
   H ——|—— OH
  HO ——|—— H
   H ——|—— OH
   H ——|—— OH
        CH₂OH
```

Fischer projection of D-Glucose

10.6 See also

- Molecular graph

- Chemical formula

- Valency interaction formula

10.7 References

[1] J. Brecher (2006). "Graphical representation of stereochemical configuration (IUPAC Recommendations 2006)" (PDF). *Pure Appl. Chem.* **78** (10): 1897–1970. doi:10.1351/pac200678101897.

10.8 External links

- The Importance of Structural Formulas

- Structural formulas

- How to get structural formulas using crystallography

Chapter 11

Stereoisomerism

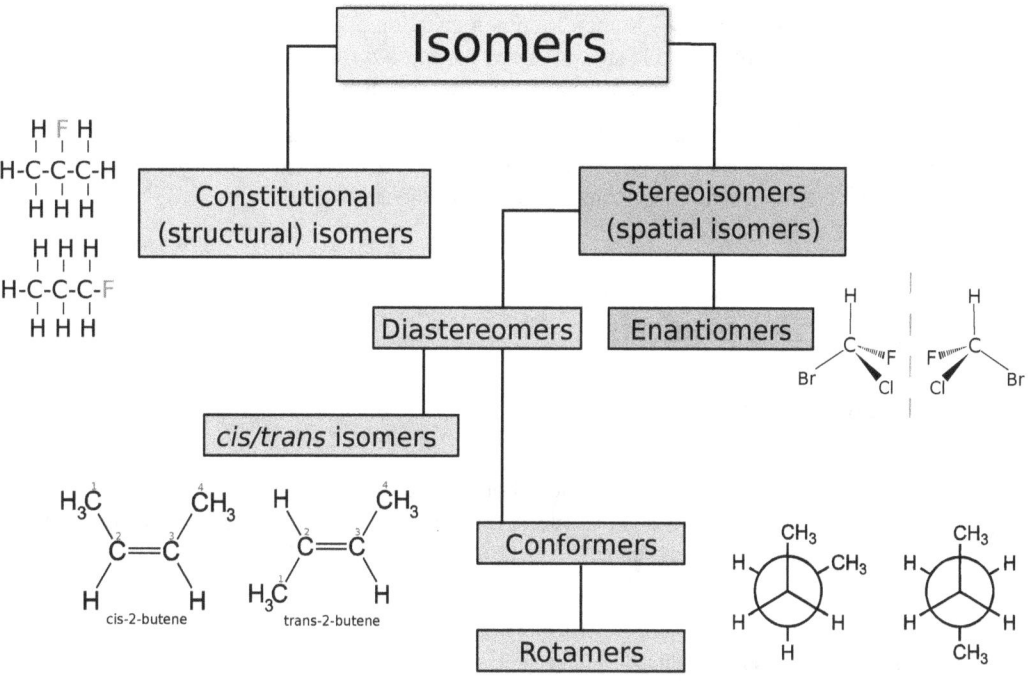

The different types of isomers. Stereochemistry focuses on stereoisomers.

Stereoisomers are isomeric molecules that have the same molecular formula and sequence of bonded atoms (constitution), but differ in the three-dimensional orientations of their atoms in space.[1][2] This contrasts with structural isomers, which share the same molecular formula, but the bond connections or their order differs. By definition, molecules that are stereoisomers of each other represent the same structural isomer.

11.1 Enantiomers

Main articles: Chirality (chemistry) and Enantiomer

Enantiomers are two stereoisomers that are related to each other by a reflection: They are mirror images of each other that are non-superimposable. Human hands are a macroscopic analog of stereoisomerism. Every stereogenic center in one has the opposite configuration in the other. Two compounds that are enantiomers of each other have the same physical properties, except for the direction in which they rotate polarized light and how they interact with different optical isomers of other compounds. As a result, different enantiomers of a compound may have substantially different biological effects. Pure enantiomers also exhibit the phenomenon of optical activity and can be separated only with the use of a chiral agent. In nature, only one enantiomer of most chiral biological compounds, such as amino acids (except glycine, which is achiral), is present.

11.2 Diastereomers

Main article: Diastereomer

Diastereomers are stereoisomers not related through a reflection operation. They are not mirror images of each other. These include meso compounds, *cis–trans* (*E-Z*) isomers, and non-enantiomeric optical isomers. Diastereomers seldom have the same physical properties. In the example shown below, the meso form of tartaric acid forms a diastereomeric pair with both levo and dextro tartaric acids, which form an enantiomeric pair.

It should be carefully noted here that the D- and L- labeling of the isomers above is not the same as the *d-* and *l-* labeling more commonly seen, explaining why these may appear reversed to those familiar with only the latter naming convention. Please refer to Chirality for more information regarding the D- and L- labels.

11.3 Cis–trans and E-Z isomerism

Main articles: Cis–trans isomerism and E-Z notation

Stereoisomerism about double bonds arises because rotation about the double bond is restricted, keeping the substituents fixed relative to each other. If the two substituents on at least one end of a double bond are the same, then there is no stereoisomer and the double bond is not a stereocenter, e.g. propene, $CH_3CH=CH_2$ where the two substituents at one end are both H.

Traditionally, double bond stereochemistry was described as either *cis* (Latin, on this side) or *trans* (Latin, across), in reference to the relative position of substituents on either side of a double bond. The simplest examples of *cis-trans* isomerism are the 1,2-disubstituted ethenes, like the dichloroethene ($C_2H_2Cl_2$) isomers shown below.

Molecule I is *cis*−1,2-dichloroethene and molecule II is *trans*−1,2-dichloroethene. Due to occasional ambiguity, IUPAC adopted a more rigorous system wherein the substituents at each end of the double bond are assigned priority based on their atomic number. If the high-priority substituents are on the same side of the bond, it is assigned Z (Ger. *zusammen*, together). If they are on opposite sides, it is E (Ger. *entgegen*, opposite). Since chlorine has a larger atomic number than hydrogen, it is the highest-priority group. Using this notation to name the above pictured molecules, molecule I is (Z)−1,2-dichloroethene and molecule II is (E)−1,2-dichloroethene. It is not the case that Z and *cis* or E and *trans* are always interchangeable. Consider the following fluoromethylpentene:

The proper name for this molecule is either *trans*−2-fluoro-3-methylpent-2-ene because the alkyl groups that form the backbone chain (i.e., methyl and ethyl) reside across the double bond from each other, or (Z)−2-fluoro-3-methylpent-2-ene because the highest-priority groups on each side of the double bond are on the same side of the double bond. Fluoro is the highest-priority group on the left side of the double bond, and ethyl is the highest-priority group on the right side of the molecule.

The terms *cis* and *trans* are also used to describe the relative position of two substituents on a ring; *cis* if on the same side, otherwise *trans*.

Dichloroethene isomers

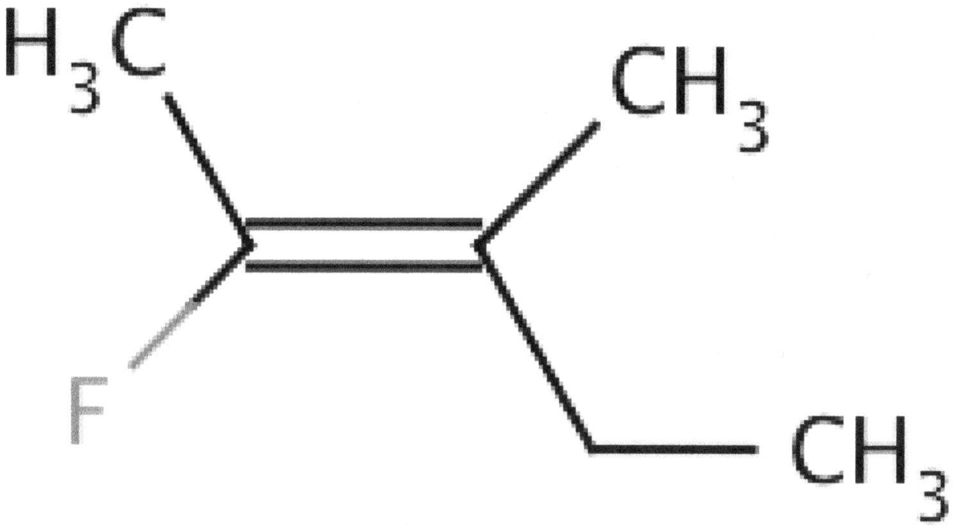

Fluoromethylpentene

11.4   Conformers

Main article: Conformational isomerism

Conformational isomerism is a form of isomerism that describes the phenomenon of molecules with the same structural formula but with different shapes due to rotations about one or more bonds. Different conformations can have different energies, can usually interconvert, and are very rarely isolatable. For example, cyclohexane can exist in a variety of different conformations including a chair conformation and a boat conformation, but, for cyclohexane itself, these can never be separated. The boat conformation represents the energy maximum on a conformational itinerary between the two equivalent chair forms; however, it does not represent the transition state for this process, because there are lower-energy pathways. There are some molecules that can be isolated in several conformations, due to the large energy barriers between different conformations. 2,2,2',2'-Tetrasubstituted biphenyls can fit into this latter category.

11.5 Atropisomers

Main article: Atropisomer

Atropisomers are stereoisomers resulting from hindered rotation about single bonds where the steric strain barrier to rotation is high enough to allow for the isolation of the conformers.

11.6 More definitions

- A **configurational stereoisomer** is a stereoisomer of a reference molecule that has the opposite configuration at a stereocenter (e.g., R- vs S- or E- vs Z-). This means that configurational isomers can be interconverted only by breaking covalent bonds to the stereocenter, for example, by inverting the configurations of some or all of the stereocenters in a compound.

11.7 Le Bel-van't Hoff rule

Le Bel-van't Hoff rule states that if n is the number of asymmetric carbon atoms then the maximum number of isomers $= 2^n$. As an example, the aldohexose D-glucose has the formula $(C \cdot H_2O)_6$, of which four of its six carbons atoms are stereogenic or asymmetric, making it one of $2^4 = 16$ possible stereoisomers.

11.8 References

[1] IUPAC, *Compendium of Chemical Terminology*, 2nd ed. (the "Gold Book") (1997). Online corrected version: (2006–) "stereoisomerism".

[2] Columbia Encyclopedia. "Stereoisomers" in *Encyclopedia.com*, n.l., **2005**, Link

Chapter 12

Stoichiometry

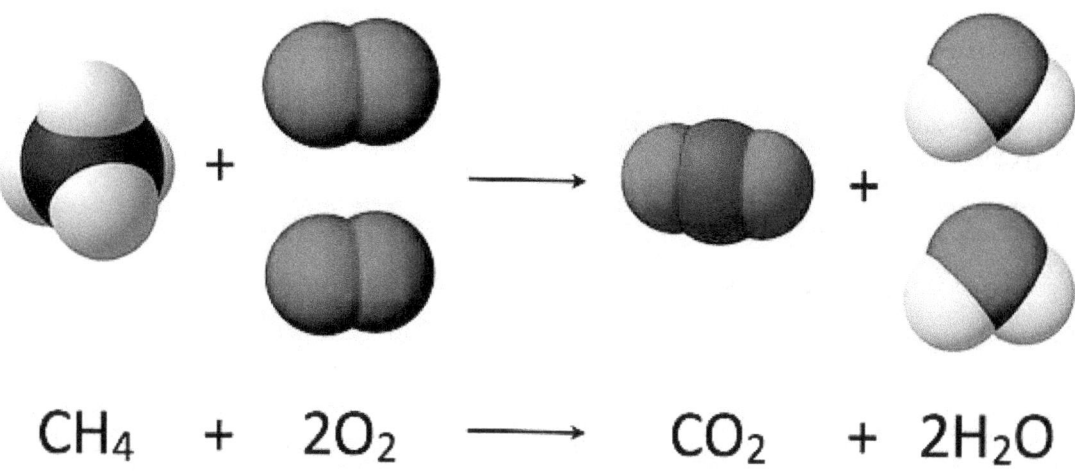

$$CH_4 \; + \; 2O_2 \; \longrightarrow \; CO_2 \; + \; 2H_2O$$

Stoichiometry /ˌstɔɪkiˈɒmɪtri/ is the calculation of relative quantities of reactants and products in chemical reactions.

Stoichiometry is founded on the law of conservation of mass where the total mass of the reactants equals the total mass of the products leading to the insight that the relations among quantities of reactants and products typically form a ratio of positive integers. This means that if the amounts of the separate reactants are known, then the amount of the product can be calculated. Conversely, if one reactant has a known quantity and the quantity of product can be empirically determined, then the amount of the other reactants can also be calculated.

As seen in the image to the right, where the balanced equation is:

CH
4 + 2 O
2 → CO
2 + 2 H
2O.

Here, one molecule of methane reacts with two molecules of oxygen gas to yield one molecule of carbon dioxide and two molecules of water. Stoichiometry measures these quantitative relationships, and is used to determine the amount of products/reactants that are produced/needed in a given reaction. Describing the quantitative relationships among substances as they participate in chemical reactions is known as *reaction stoichiometry*. In the example above, reaction stoichiometry measures the relationship between the methane and oxygen as they react to form carbon dioxide and water.

Because of the well known relationship of moles to atomic weights, the ratios that are arrived at by stoichiometry can be used to determine quantities by weight in a reaction described by a balanced equation. This is called *composition stoichiometry*.

Gas stoichiometry deals with reactions involving gases, where the gases are at a known temperature, pressure, and volume and can be assumed to be ideal gases. For gases, the volume ratio is ideally the same by the ideal gas law, but the mass ratio of a single reaction has to be calculated from the molecular masses of the reactants and products. In practice, due to the existence of isotopes, molar masses are used instead when calculating the mass ratio.

12.1 Etymology

The term *stoichiometry* was first used by Jeremias Benjamin Richter in 1792 when the first volume of Richter's *Stoichiometry or the Art of Measuring the Chemical Elements* was published. The term is derived from the Greek words στοιχεῖον *stoicheion* "element" and μέτρον *metron* "measure". In patristic Greek, the word *Stoichiometria* was used by Nicephorus to refer to the number of line counts of the canonical New Testament and some of the Apocrypha.

12.2 Definition

A *stoichiometric amount* or *stoichiometric ratio* of a reagent is the optimum amount or ratio where, assuming that the reaction proceeds to completion:

1. All of the reagent is consumed

2. There is no deficiency of the reagent

3. There is no excess of the reagent.

Stoichiometry rests upon the very basic laws that help to understand it better, i.e., law of conservation of mass, the law of definite proportions (i.e., the law of constant composition), the law of multiple proportions and the law of reciprocal proportions. In general, chemical reactions combine in definite ratios of chemicals. Since chemical reactions can neither create nor destroy matter, nor transmute one element into another, the amount of each element must be the same throughout the overall reaction. For example, the number of atoms of a given element X on the reactant side must equal the number of atoms of that element on the product side, whether or not all of those atoms are actually involved in a reaction.

Chemical reactions, as macroscopic unit operations, consist of simply a very large number of elementary reactions, where a single molecule reacts with another molecule. As the reacting molecules (or moieties) consist of a definite set of atoms in an integer ratio, the ratio between reactants in a complete reaction is also in integer ratio. A reaction may consume more than one molecule, and the **stoichiometric number** counts this number, defined as positive for products (added) and negative for reactants (removed).[1]

Different elements have a different atomic mass, and as collections of single atoms, molecules have a definite molar mass, measured with the unit mole (6.02×10^{23} individual molecules, Avogadro's constant). By definition, carbon-12 has a molar mass of 12 g/mol. Thus, to calculate the stoichiometry by mass, the number of molecules required for each reactant is expressed in moles and multiplied by the molar mass of each to give the mass of each reactant per mole of reaction. The mass ratios can be calculated by dividing each by the total in the whole reaction.

Elements in their natural state are mixtures of isotopes of differing mass, thus atomic masses and thus molar masses are not exactly integers. For instance, instead of an exact 14:3 proportion, 17.04 kg of ammonia consists of 14.01 kg of nitrogen and 3×1.01 kg of hydrogen, because natural nitrogen includes a small amount of nitrogen-15, and natural hydrogen includes hydrogen-2 (deuterium).

A **stoichiometric reactant** is a reactant that is consumed in a reaction, as opposed to a catalytic reactant, which is not consumed in the overall reaction because it reacts in one step and is regenerated in another step.

12.3   Converting grams to moles

Stoichiometry is not only used to balance chemical equations but also used in conversions, i.e., converting from grams to moles, or from grams to milliliters. For example, to find the amount of NaCl (sodium chloride) in 2.00 g, one would do the following:

$$\frac{2.00 \text{ g NaCl}}{58.44 \text{ g NaCl mol}^{-1}} = 0.034 \text{ mol}$$

In the above example, when written out in fraction form, the units of grams form a multiplicative identity, which is equivalent to one (g/g=1), with the resulting amount in moles (the unit that was needed), as shown in the following equation,

$$\left(\frac{2.00 \text{ g NaCl}}{1}\right)\left(\frac{1 \text{ mol NaCl}}{58.44 \text{ g NaCl}}\right) = 0.034 \text{ mol}$$

12.4   Molar proportion

Stoichiometry is often used to balance chemical equations (reaction stoichiometry). For example, the two diatomic gases, hydrogen and oxygen, can combine to form a liquid, water, in an exothermic reaction, as described by the following equation:

2 H
2 + O
2 → 2 H
2O

Reaction stoichiometry describes the 2:1:2 ratio of hydrogen, oxygen, and water molecules in the above equation.

The molar ratio allows for conversion between moles of one substance and moles of another. For example, in the reaction

2 CH
3OH + 3 O
2 → 2 CO
2 + 4 H
2O

the amount of water that will be produced by the combustion of 0.27 moles of CH
3OH is obtained using the molar ratio between CH
3OH and H
2O of 2 to 4.

$$\left(\frac{0.27 \text{ mol CH}_3\text{OH}}{1}\right)\left(\frac{4 \text{ mol H}_2\text{O}}{2 \text{ mol CH}_3\text{OH}}\right) = 0.54 \text{ mol H}_2\text{O}$$

The term stoichiometry is also often used for the molar proportions of elements in stoichiometric compounds (composition stoichiometry). For example, the stoichiometry of hydrogen and oxygen in H_2O is 2:1. In stoichiometric compounds, the molar proportions are whole numbers.

12.5 Determining amount of product

Stoichiometry can also be used to find the quantity of a product yielded by a reaction. If a piece of solid copper (Cu) were added to an aqueous solution of silver nitrate ($AgNO_3$), the silver (Ag) would be replaced in a single displacement reaction forming aqueous copper(II) nitrate ($Cu(NO_3)_2$) and solid silver. How much silver is produced if 16.00 grams of Cu is added to the solution of excess silver nitrate?

The following steps would be used:

1. Write and Balance the Equation

2. Mass to Mole: Convert g Cu to moles Cu

3. Mole Ratio: Convert moles of Cu to moles of Ag produced

4. Mole to Mass: Convert moles Ag to grams of Ag produced

The complete balanced equation would be:

$$Cu + 2\,AgNO_3 \rightarrow Cu(NO_3)_2 + 2\,Ag$$

For the mass to mole step, the mass of copper (16.00 g) would be converted to moles of copper by dividing the mass of copper by its molecular mass: 63.55 g/mol.

$$\left(\frac{16.00\text{ g Cu}}{1}\right)\left(\frac{1\text{ mol Cu}}{63.55\text{ g Cu}}\right) = 0.2518\text{ Cu mol}$$

Now that the amount of Cu in moles (0.2518) is found, we can set up the mole ratio. This is found by looking at the coefficients in the balanced equation: Cu and Ag are in a 1:2 ratio.

$$\left(\frac{0.2518\text{ mol Cu}}{1}\right)\left(\frac{2\text{ mol Ag}}{1\text{ mol Cu}}\right) = 0.5036\text{ Ag mol}$$

Now that the moles of Ag produced is known to be 0.5036 mol, we convert this amount to grams of Ag produced to come to the final answer:

$$\left(\frac{0.5036\text{ mol Ag}}{1}\right)\left(\frac{107.87\text{ g Ag}}{1\text{ mol Ag}}\right) = 54.32\text{ Ag g}$$

This set of calculations can be further condensed into a single step:

$$m_{Ag} = \left(\frac{16.00\text{ g Cu}}{1}\right)\left(\frac{1\text{ mol Cu}}{63.55\text{ g Cu}}\right)\left(\frac{2\text{ mol Ag}}{1\text{ mol Cu}}\right)\left(\frac{107.87\text{ g Ag}}{1\text{ mol Ag}}\right) = 54.32\text{ g}$$

12.5.1 Further examples

For propane (C_3H_8) reacting with oxygen gas (O_2), the balanced chemical equation is:

C
3H
8 + 5 O
2 → 3 CO
2 + 4 H
2O

The mass of water formed if 120 g of propane (C_3H_8) is burned in excess oxygen is then

$$m_{H_2O} = \left(\frac{120.\ g\ C_3H_8}{1} \right) \left(\frac{1\ mol\ C_3H_8}{44.09\ g\ C_3H_8} \right) \left(\frac{4\ mol\ H_2O}{1\ mol\ C_3H_8} \right) \left(\frac{18.02\ g\ H_2O}{1\ mol\ H_2O} \right) = 196\ g$$

12.6   Stoichiometric ratio

Stoichiometry is also used to find the right amount of one reactant to "completely" react with the other reactant in a chemical reaction - that is, the stoichiometric amounts that would result in no leftover reactants when the reaction takes place. An example is shown below using the thermite reaction,

Fe
2O
3 + 2 Al → Al
2O
3 + 2 Fe

This equation shows that 1 mole of iron(III) oxide and 2 moles of aluminum will produce 1 mole of aluminium oxide and 2 moles of iron. So, to completely react with 85.0 g of iron(III) oxide (0.532 mol), 28.7 g (1.06 mol) of aluminium are needed.

$$m_{Al} = \left(\frac{85.0\ g\ Fe_2O_3}{1} \right) \left(\frac{1\ mol\ Fe_2O_3}{159.7\ g\ Fe_2O_3} \right) \left(\frac{2\ mol\ Al}{1\ mol\ Fe_2O_3} \right) \left(\frac{26.98\ g\ Al}{1\ mol\ Al} \right) = 28.7\ g$$

12.7   Limiting reagent and percent yield

Main articles: Limiting reagent and Yield (chemistry)

The limiting reagent is the reagent that limits the amount of product that can be formed and is completely consumed during the reaction. The excess reactant is the reactant that is left over once the reaction has stopped due to the limiting reactant.

Consider the equation of roasting lead(II) sulfide (PbS) in oxygen (O_2) to produce lead(II) oxide (PbO) and sulfur dioxide (SO_2):

2 PbS + 3 O
2 → 2 PbO + 2 SO
2

To determine the theoretical yield of lead(II) oxide if 200.0 g of lead(II) sulfide and 200.0 grams of oxygen are heated in an open container:

$$m_{PbO} = \left(\frac{200.0 \text{ g PbS}}{1}\right)\left(\frac{1 \text{ mol PbS}}{239.27 \text{ g PbS}}\right)\left(\frac{2 \text{ mol PbO}}{2 \text{ mol PbS}}\right)\left(\frac{223.2 \text{ g PbO}}{1 \text{ mol PbO}}\right) = 186.6 \text{ g}$$

$$m_{PbO} = \left(\frac{200.0 \text{ g O}_2}{1}\right)\left(\frac{1 \text{ mol O}_2}{32.00 \text{ g O}_2}\right)\left(\frac{2 \text{ mol PbO}}{3 \text{ mol O}_2}\right)\left(\frac{223.2 \text{ g PbO}}{1 \text{ mol PbO}}\right) = 930.0 \text{ g}$$

Because a lesser amount of PbO is produced for the 200.0 g of PbS, it is clear that PbS is the limiting reagent.

In reality, the actual yield is not the same as the stoichiometrically-calculated theoretical yield. Percent yield, then, is expressed in the following equation:

$$\text{percent yield} = \frac{\text{actual yield}}{\text{theoretical yield}} \times 100$$

If 170.0 g of lead(II) oxide is obtained, then the percent yield would be calculated as follows:

$$\text{percent yield} = \frac{170.0 \text{ g PbO}}{186.6 \text{ g PbO}} \times 100 = 91.12\%$$

12.7.1 Example

Consider the following reaction, in which iron(III) chloride reacts with hydrogen sulfide to produce iron(III) sulfide and hydrogen chloride:

2 FeCl

3 + 3 H

2S → Fe

2S

3 + 6 HCl

Suppose 90.0 g of $FeCl_3$ reacts with 52.0 g of H_2 S. To find the limiting reagent and the mass of HCl produced by the reaction, we could set up the following equations:

$$m_{HCl} = \left(\frac{90.0 \text{ g FeCl}_3}{1}\right)\left(\frac{1 \text{ mol FeCl}_3}{162 \text{ g FeCl}_3}\right)\left(\frac{6 \text{ mol HCl}}{2 \text{ mol FeCl}_3}\right)\left(\frac{36.5 \text{ g HCl}}{1 \text{ mol HCl}}\right) = 60.8 \text{ g}$$

$$m_{HCl} = \left(\frac{52.0 \text{ g H}_2\text{S}}{1}\right)\left(\frac{1 \text{ mol H}_2\text{S}}{34.1 \text{ g H}_2\text{S}}\right)\left(\frac{6 \text{ mol HCl}}{3 \text{ mol H}_2\text{S}}\right)\left(\frac{36.5 \text{ g HCl}}{1 \text{ mol HCl}}\right) = 111 \text{ g}$$

Thus, the limiting reagent is $FeCl_3$ and the amount of HCl produced is 60.8 g.

To find what mass of excess reagent (H_2S) remains after the reaction, we would set up the calculation to find out how much H_2S reacts completely with the 90.0 g $FeCl_3$:

$$m_{H_2S} = \left(\frac{90.0 \text{ g FeCl}_3}{1}\right)\left(\frac{1 \text{ mol FeCl}_3}{162 \text{ g FeCl}_3}\right)\left(\frac{3 \text{ mol H}_2\text{S}}{2 \text{ mol FeCl}_3}\right)\left(\frac{34.1 \text{ g H}_2\text{S}}{1 \text{ mol H}_2\text{S}}\right) = 28.4 \text{ g reacted}$$

By subtracting this amount from the original amount of H_2S, we can come to the answer:

52.0 g H_2S − 28.4 g H_2S = 23.6 g H_2S excess

12.8   Different stoichiometries in competing reactions

Often, more than one reaction is possible given the same starting materials. The reactions may differ in their stoichiometry. For example, the methylation of benzene (C_6H_6), through a Friedel-Crafts reaction using $AlCl_3$ as a catalyst, may produce singly methylated ($C_6H_5CH_3$,) doubly methylated ($C_6H_4(CH_3)_2$), or still more highly methylated ($C_6H_{6-n}(CH_3)_n$) products, as shown in the following example,

$$C_6H_6 + CH_3Cl \rightarrow C_6H_5CH_3 + HCl$$

$$C_6H_6 + 2\,CH_3Cl \rightarrow C_6H_4(CH_3)_2 + 2\,HCl$$

$$C_6H_6 + n\,CH_3Cl \rightarrow C_6H_{6\text{-}n}(CH_3)_n + n\,HCl$$

In this example, which reaction takes place is controlled in part by the relative concentrations of the reactants.

12.9   Stoichiometric coefficient

In lay terms, the *stoichiometric coefficient* (or *stoichiometric number* in the IUPAC nomenclature[2]) of any given component is the number of molecules that participate in the reaction as written.

For example, in the reaction $CH_4 + 2\,O_2 \rightarrow CO_2 + 2\,H_2O$, the stoichiometric coefficient of CH_4 is -1, the stoichiometric coefficient of O_2 is -2, for CO_2 it would be $+1$ and for H_2O it is $+2$.

In more technically precise terms, the stoichiometric coefficient in a chemical reaction system of the *i–th* component is defined as

$$\nu_i = \frac{\Delta N_i}{\Delta \xi}$$

or

$$\Delta N_i = \nu_i \Delta \xi$$

where N_i is the number of molecules of i, and ξ is the progress variable or **extent of reaction**.[3]

> The **extent of reaction** ξ can be regarded as [the amount of] a real (or hypothetical) product, one molecule of which produced each time the reaction event occurs. It is the extensive quantity describing the progress of a chemical reaction equal to the number of chemical transformations, as indicated by the reaction equation on a molecular scale, divided by the Avogadro constant (in essence, it is the amount of chemical transformations). The change in the extent of reaction is given by $d\xi = dn_B/\nu_B$, where ν_B is the stoichiometric number of any reaction entity B (reactant or product) and n_B is the corresponding amount.[4]

The stoichiometric coefficient ν_i represents the degree to which a chemical species participates in a reaction. The convention is to assign negative coefficients to *reactants* (which are consumed) and positive ones to *products*. However, any reaction may be viewed as "going" in the reverse direction, and all the coefficients then change sign (as does the free energy). Whether a reaction actually *will* go in the arbitrarily selected forward direction or not depends on the amounts of the substances present at any given time, which determines the kinetics and thermodynamics, i.e., whether equilibrium lies to the *right* or the *left*.

In reaction mechanisms, stoichiometric coefficients for each step are always integers, since elementary reactions always involve whole molecules. If one uses a composite representation of an "overall" reaction, some may be rational fractions. There are often chemical species present that do not participate in a reaction; their stoichiometric coefficients are therefore zero. Any chemical species that is regenerated, such as a catalyst, also has a stoichiometric coefficient of zero.

The simplest possible case is an isomerism

$$A \rightarrow B$$

in which $\nu B = 1$ since one molecule of B is produced each time the reaction occurs, while $\nu A = -1$ since one molecule of A is necessarily consumed. In any chemical reaction, not only is the total mass conserved but also the numbers of atoms of each kind are conserved, and this imposes corresponding constraints on possible values for the stoichiometric coefficients.

There are usually multiple reactions proceeding simultaneously in any natural reaction system, including those in biology. Since any chemical component can participate in several reactions simultaneously, the stoichiometric coefficient of the i–th component in the k–th reaction is defined as

$$\nu_{ik} = \frac{\partial N_i}{\partial \xi_k}$$

so that the total (differential) change in the amount of the i–th component is

$$dN_i = \sum_k \nu_{ik} d\xi_k.$$

Extents of reaction provide the clearest and most explicit way of representing compositional change, although they are not yet widely used.

With complex reaction systems, it is often useful to consider both the representation of a reaction system in terms of the amounts of the chemicals present { N_i } (state variables), and the representation in terms of the actual compositional degrees of freedom, as expressed by the extents of reaction { ξ_k }. The transformation from a vector expressing the extents to a vector expressing the amounts uses a rectangular matrix whose elements are the stoichiometric coefficients [ν_{ik}].

The maximum and minimum for any ξ_k occur whenever the first of the reactants is depleted for the forward reaction; or the first of the "products" is depleted if the reaction as viewed as being pushed in the reverse direction. This is a purely kinematic restriction on the reaction simplex, a hyperplane in composition space, or N-space, whose dimensionality equals the number of *linearly-independent* chemical reactions. This is necessarily less than the number of chemical components, since each reaction manifests a relation between at least two chemicals. The accessible region of the hyperplane depends on the amounts of each chemical species actually present, a contingent fact. Different such amounts can even generate different hyperplanes, all sharing the same algebraic stoichiometry.

In accord with the principles of chemical kinetics and thermodynamic equilibrium, every chemical reaction is *reversible*, at least to some degree, so that each equilibrium point must be an interior point of the simplex. As a consequence, extrema for the ξ's will not occur unless an experimental system is prepared with zero initial amounts of some products.

The number of *physically*-independent reactions can be even greater than the number of chemical components, and depends on the various reaction mechanisms. For example, there may be two (or more) reaction *paths* for the isomerism above. The reaction may occur by itself, but faster and with different intermediates, in the presence of a catalyst.

The (dimensionless) "units" may be taken to be molecules or moles. Moles are most commonly used, but it is more suggestive to picture incremental chemical reactions in terms of molecules. The N's and ξ's are reduced to molar units by dividing by Avogadro's number. While dimensional mass units may be used, the comments about integers are then no longer applicable.

12.10 Stoichiometry matrix

Main article: Chemical reaction network theory

In complex reactions, stoichiometries are often represented in a more compact form called the stoichiometry matrix. The stoichiometry matrix is denoted by the symbol $\mathbf{N}$.

If a reaction network has n reactions and m participating molecular species then the stoichiometry matrix will have corresponding m rows and n columns.

For example, consider the system of reactions shown below:

$$S_1 \rightarrow S_2$$
$$5\,S_3 + S_2 \rightarrow 4\,S_3 + 2\,S_2$$
$$S_3 \rightarrow S_4$$
$$S_4 \rightarrow S_5.$$

This systems comprises four reactions and five different molecular species. The stoichiometry matrix for this system can be written as:

$$\mathbf{N} = \begin{bmatrix} -1 & 0 & 0 & 0 \\ 1 & 1 & 0 & 0 \\ 0 & -1 & -1 & 0 \\ 0 & 0 & 1 & -1 \\ 0 & 0 & 0 & 1 \end{bmatrix}$$

where the rows correspond to S_1, S_2, S_3, S_4 and S_5, respectively. Note that the process of converting a reaction scheme into a stoichiometry matrix can be a lossy transformation, for example, the stoichiometries in the second reaction simplify when included in the matrix. This means that it is not always possible to recover the original reaction scheme from a stoichiometry matrix.

Often the stoichiometry matrix is combined with the rate vector, $\mathbf{v}$, and the species vector, $\mathbf{S}$ to form a compact equation describing the rates of change of the molecular species:

$$\frac{d\mathbf{S}}{dt} = \mathbf{N} \cdot \mathbf{v}.$$

12.11   Gas stoichiometry

Gas stoichiometry is the quantitative relationship (ratio) between reactants and products in a chemical reaction with reactions that produce gases. Gas stoichiometry applies when the gases produced are assumed to be ideal, and the temperature, pressure, and volume of the gases are all known. The ideal gas law is used for these calculations. Often, but not always, the standard temperature and pressure (STP) are taken as 0 °C and 1 bar and used as the conditions for gas stoichiometric calculations.

Gas stoichiometry calculations solve for the unknown volume or mass of a gaseous product or reactant. For example, if we wanted to calculate the volume of gaseous NO_2 produced from the combustion of 100 g of NH_3, by the reaction:

$$4\,NH_{3(g)} + 7\,O_{2(g)} \rightarrow 4\,NO_{2(g)} + 6\,H_2O_{(l)}$$

we would carry out the following calculations:

$$100\,\text{g}\,NH_3 \cdot \frac{1\,\text{mol}\,NH_3}{17.034\,\text{g}\,NH_3} = 5.871\,\text{mol}\,NH_3$$

There is a 1:1 molar ratio of NH_3 to NO_2 in the above balanced combustion reaction, so 5.871 mol of NO_2 will be formed. We will employ the ideal gas law to solve for the volume at 0 °C (273.15 K) and 1 atmosphere using the gas law constant of R = 0.08206 L·atm·K^{-1}·mol^{-1} :

$$PV = nRT$$
$$V = \frac{nRT}{P}$$
$$= \frac{5.871 \cdot 0.08206 \cdot 273.15}{1}$$
$$= 131.597 \, \text{L NO}_2$$

Gas stoichiometry often involves having to know the molar mass of a gas, given the density of that gas. The ideal gas law can be re-arranged to obtain a relation between the density and the molar mass of an ideal gas:

$$\rho = \tfrac{m}{V} \text{ and } n = \tfrac{m}{M}$$

and thus:

$$\rho = \frac{MP}{RT}$$

where:

- P = absolute gas pressure

- V = gas volume

- n = amount (measured in moles)

- R = universal ideal gas law constant

- T = absolute gas temperature

- ϱ = gas density at T and P

- m = mass of gas

- M = molar mass of gas

12.12   Stoichiometric air-to-fuel ratios of common fuels

See also: Air–fuel ratio and Combustion

In the combustion reaction, oxygen reacts with the fuel, and the point where exactly all oxygen is consumed and all fuel burned is defined as the stoichiometric point. With more oxygen (overstoichiometric combustion), some of it stays unreacted. Likewise, if the combustion is incomplete due to lack of sufficient oxygen, fuel remains unreacted. (Unreacted fuel may also remain because of slow combustion or insufficient mixing of fuel and oxygen - this is not due to stoichiometry. Different hydrocarbon fuels have different contents of carbon, hydrogen and other elements, thus their stoichiometry varies.

Gasoline engines can run at stoichiometric air-to-fuel ratio, because gasoline is quite volatile and is mixed (sprayed or carburetted) with the air prior to ignition. Diesel engines, in contrast, run lean, with more air available than simple stoichiometry would require. Diesel fuel is less volatile and is effectively burned as it is injected, leaving less time for evaporation and mixing. Thus, it would form soot (black smoke) at stoichiometric ratio.

12.13   References

[1] "stoichiometric number,". *iupac.org.*

[2] IUPAC. Compendium of Chemical Terminology, 2nd ed. (the "Gold Book"). Compiled by A. D. McNaught and A. Wilkinson. Blackwell Scientific Publications, Oxford (1997). XML on-line corrected version: http://goldbook.iupac.org (2006-) created by M. Nic, J. Jirat, B. Kosata; updates compiled by A. Jenkins. ISBN 0-9678550-9-8. doi:10.1351/goldbook. Entry: "stoichiometric number".

[3] Prigogine & Defay, p. 18; Prigogine, pp. 4–7; Guggenheim, p. 37 & 62

[4] IUPAC Gold Book of Chemical Terminology, last visited May 4, 2015.

[5] John B. Heywood: "Internal Combustion Engine Fundamentals page 915", 1988

[6] North American Mfg. Co.: "North American Combustion Handbook", 1952

- Zumdahl, Steven S. *Chemical Principles.* Houghton Mifflin, New York, 2005, pp 148–150.

- Internal Combustion Engine Fundamentals, John B. Heywood

12.14   External links

- Engine Combustion primer from the University of Plymouth

- Free Stoichiometry Tutorials from Carnegie Mellon's ChemCollective

- Stoichiometry Add-In for Microsoft Excel for calculation of molecular weights, reaction coëfficients and stoichiometry.

- Reaction Stoichiometry Calculator a comprehensive free online reaction stoichiometry calculator.

- Stoichiometry Calculator a unit conversion calculator for individual compound stoichiometry.

Chapter 13

Spectroscopy

Spectroscopy /spɛkˈtrɒskəpi/ is the study of the interaction between matter and electromagnetic radiation.[1][2] Historically, spectroscopy originated through the study of visible light dispersed according to its wavelength, by a prism. Later the concept was expanded greatly to comprise any interaction with radiative energy as a function of its wavelength or frequency. Spectroscopic data is often represented by a spectrum, a plot of the response of interest as a function of wavelength or frequency.

13.1 Introduction

Spectroscopy and **spectrography** are terms used to refer to the measurement of radiation intensity as a function of wavelength and are often used to describe experimental spectroscopic methods. Spectral measurement devices are referred to as spectrometers, spectrophotometers, spectrographs or spectral analyzers.

Daily observations of color can be related to spectroscopy. Neon lighting is a direct application of atomic spectroscopy. Neon and other noble gases have characteristic emission frequencies (colors). Neon lamps use collision of electrons with the gas to excite these emissions. Inks, dyes and paints include chemical compounds selected for their spectral characteristics in order to generate specific colors and hues. A commonly encountered molecular spectrum is that of nitrogen dioxide. Gaseous nitrogen dioxide has a characteristic red absorption feature, and this gives air polluted with nitrogen dioxide a reddish brown color. Rayleigh scattering is a spectroscopic scattering phenomenon that accounts for the color of the sky.

Spectroscopic studies were central to the development of quantum mechanics and included Max Planck's explanation of blackbody radiation, Albert Einstein's explanation of the photoelectric effect and Niels Bohr's explanation of atomic structure and spectra. Spectroscopy is used in physical and analytical chemistry because atoms and molecules have unique spectra. As a result, these spectra can be used to detect, identify and quantify information about the atoms and molecules. Spectroscopy is also used in astronomy and remote sensing on earth. Most research telescopes have spectrographs. The measured spectra are used to determine the chemical composition and physical properties of astronomical objects (such as their temperature and velocity).

13.2 Theory

One of the central concepts in spectroscopy is a resonance and its corresponding resonant frequency. Resonances were first characterized in mechanical systems such as pendulums. Mechanical systems that vibrate or oscillate will experience large amplitude oscillations when they are driven at their resonant frequency. A plot of amplitude vs. excitation frequency will have a peak centered at the resonance frequency. This plot is one type of spectrum, with the peak often referred to as a spectral line, and most spectral lines have a similar appearance.

In quantum mechanical systems, the analogous resonance is a coupling of two quantum mechanical stationary states of one

Analysis of white light by dispersing it with a prism is an example of spectroscopy.

system, such as an atom, via an oscillatory source of energy such as a photon. The coupling of the two states is strongest when the energy of the source matches the energy difference between the two states. The energy (E) of a photon is related to its frequency (ν) by $E = h\nu$ where h is Planck's constant, and so a spectrum of the system response vs. photon frequency will peak at the resonant frequency or energy. Particles such as electrons and neutrons have a comparable relationship, the de Broglie relations, between their kinetic energy and their wavelength and frequency and therefore can also excite resonant interactions.

Spectra of atoms and molecules often consist of a series of spectral lines, each one representing a resonance between two different quantum states. The explanation of these series, and the spectral patterns associated with them, were one of the experimental enigmas that drove the development and acceptance of quantum mechanics. The hydrogen spectral series in particular was first successfully explained by the Rutherford-Bohr quantum model of the hydrogen atom. In some cases spectral lines are well separated and distinguishable, but spectral lines can also overlap and appear to be a single transition if the density of energy states is high enough. Named series of lines include the principal, sharp, diffuse and fundamental series.

13.3 Classification of methods

Spectroscopy is a sufficiently broad field that many sub-disciplines exist, each with numerous implementations of specific spectroscopic techniques. The various implementations and techniques can be classified in several ways.

13.3.1 Type of radiative energy

Types of spectroscopy are distinguished by the type of radiative energy involved in the interaction. In many applications, the spectrum is determined by measuring changes in the intensity or frequency of this energy. The types of radiative energy studied include:

- Electromagnetic radiation was the first source of energy used for spectroscopic studies. Techniques that employ electromagnetic radiation are typically classified by the wavelength region of the spectrum and include microwave, terahertz, infrared, near infrared, visible and ultraviolet, x-ray and gamma spectroscopy.

- Particles, due to their de Broglie wavelength, can also be a source of radiative energy and both electrons and neutrons are commonly used. For a particle, its kinetic energy determines its wavelength.

- Acoustic spectroscopy involves radiated pressure waves.

- Mechanical methods can be employed to impart radiating energy, similar to acoustic waves, to solid materials.

13.3.2 Nature of the interaction

Types of spectroscopy can also be distinguished by the nature of the interaction between the energy and the material. These interactions include:[1]

- Absorption occurs when energy from the radiative source is absorbed by the material. Absorption is often determined by measuring the fraction of energy transmitted through the material; absorption will decrease the transmitted portion.

- Emission indicates that radiative energy is released by the material. A material's blackbody spectrum is a spontaneous emission spectrum determined by its temperature; this features can be measured in the infrared by instruments such as the Atmospheric Emitted Radiance Interferometer (AERI).[3] Emission can also be induced by other sources of energy such as flames or sparks or electromagnetic radiation in the case of fluorescence.

- Elastic scattering and reflection spectroscopy determine how incident radiation is reflected or scattered by a material. Crystallography employs the scattering of high energy radiation, such as x-rays and electrons, to examine the arrangement of atoms in proteins and solid crystals.

- Impedance spectroscopy studies the ability of a medium to impede or slow the transmittance of energy. For optical applications, this is characterized by the index of refraction.

- Inelastic scattering phenomena involve an exchange of energy between the radiation and the matter that shifts the wavelength of the scattered radiation. These include Raman and Compton scattering.

- Coherent or resonance spectroscopy are techniques where the radiative energy couples two quantum states of the material in a coherent interaction that is sustained by the radiating field. The coherence can be disrupted by other interactions, such as particle collisions and energy transfer, and so often require high intensity radiation to be sustained. Nuclear magnetic resonance (NMR) spectroscopy is a widely used resonance method and ultrafast laser methods are also now possible in the infrared and visible spectral regions.

13.3.3   Type of material

Spectroscopic studies are designed so that the radiant energy interacts with specific types of matter.

Atoms

Atomic spectroscopy was the first application of spectroscopy developed. Atomic absorption spectroscopy (AAS) and atomic emission spectroscopy (AES) involve visible and ultraviolet light. These absorptions and emissions, often referred to as atomic spectral lines, are due to electronic transitions of outer shell electrons as they rise and fall from one electron orbit to another. Atoms also have distinct x-ray spectra that are attributable to the excitation of inner shell electrons to excited states.

Atoms of different elements have distinct spectra and therefore atomic spectroscopy allows for the identification and quantitation of a sample's elemental composition. Robert Bunsen and Gustav Kirchhoff discovered new elements by observing their emission spectra. Atomic absorption lines are observed in the solar spectrum and referred to as Fraunhofer lines after their discoverer. A comprehensive explanation of the hydrogen spectrum was an early success of quantum mechanics and explained the Lamb shift observed in the hydrogen spectrum led to the development of quantum electrodynamics.

Modern implementations of atomic spectroscopy for studying visible and ultraviolet transitions include flame emission spectroscopy, inductively coupled plasma atomic emission spectroscopy, glow discharge spectroscopy, microwave induced plasma spectroscopy, and spark or arc emission spectroscopy. Techniques for studying x-ray spectra include X-ray spectroscopy and X-ray fluorescence (XRF).

Molecules

The combination of atoms into molecules leads to the creation of unique types of energetic states and therefore unique spectra of the transitions between these states. Molecular spectra can be obtained due to electron spin states (electron paramagnetic resonance), molecular rotations, molecular vibration and electronic states. Rotations are collective motions of the atomic nuclei and typically lead to spectra in the microwave and millimeter-wave spectral regions; rotational spectroscopy and microwave spectroscopy are synonymous. Vibrations are relative motions of the atomic nuclei and are studied by both infrared and Raman spectroscopy. Electronic excitations are studied using visible and ultraviolet spectroscopy as well as fluorescence spectroscopy.

Studies in molecular spectroscopy led to the development of the first maser and contributed to the subsequent development of the laser.

Crystals and extended materials

The combination of atoms or molecules into crystals or other extended forms leads to the creation of additional energetic states. These states are numerous and therefore have a high density of states. This high density often makes the spectra weaker and less distinct, i.e., broader. For instance, blackbody radiation is due to the thermal motions of atoms and molecules within a material. Acoustic and mechanical responses are due to collective motions as well.

Pure crystals, though, can have distinct spectral transitions and the crystal arrangement also has an effect on the observed molecular spectra. The regular lattice structure of crystals also scatters x-rays, electrons or neutrons allowing for crystallographic studies.

Nuclei

Nuclei also have distinct energy states that are widely separated and lead to gamma ray spectra. Distinct nuclear spin states can have their energy separated by a magnetic field, and this allows for NMR spectroscopy.

13.4 Other types

Other types of spectroscopy are distinguished by specific applications or implementations:

- Acoustic resonance spectroscopy is based on sound waves primarily in the audible and ultrasonic regions

- Auger spectroscopy is a method used to study surfaces of materials on a micro-scale. It is often used in connection with electron microscopy.

- Cavity ring down spectroscopy

- Circular Dichroism spectroscopy

- Coherent anti-Stokes Raman spectroscopy (CARS) is a recent technique that has high sensitivity and powerful applications for *in vivo* spectroscopy and imaging.[4]

- Cold vapour atomic fluorescence spectroscopy

- Correlation spectroscopy encompasses several types of two-dimensional NMR spectroscopy.

- Deep-level transient spectroscopy measures concentration and analyzes parameters of electrically active defects in semiconducting materials

- Dual polarisation interferometry measures the real and imaginary components of the complex refractive index

- Electron phenomenological spectroscopy measures physicochemical properties and characteristics of electronic structure of multicomponent and complex molecular systems.

- EPR spectroscopy

- Force spectroscopy

- Fourier transform spectroscopy is an efficient method for processing spectra data obtained using interferometers. Fourier transform infrared spectroscopy (FTIR) is a common implementation of infrared spectroscopy. NMR also employs Fourier transforms.

- Hadron spectroscopy studies the energy/mass spectrum of hadrons according to spin, parity, and other particle properties. Baryon spectroscopy and meson spectroscopy are both types of hadron spectroscopy.

- Hyperspectral imaging is a method to create a complete picture of the environment or various objects, each pixel containing a full visible, VNIR, NIR, or infrared spectrum.

- Inelastic electron tunneling spectroscopy (IETS) uses the changes in current due to inelastic electron-vibration interaction at specific energies that can also measure optically forbidden transitions.

- Inelastic neutron scattering is similar to Raman spectroscopy, but uses neutrons instead of photons.

- Laser-Induced Breakdown Spectroscopy (LIBS), also called Laser-induced plasma spectrometry (LIPS)

- Laser spectroscopy uses tunable lasers[5] and other types of coherent emission sources, such as optical parametric oscillators,[6] for selective excitation of atomic or molecular species.

- Mass spectroscopy is an historical term used to refer to mass spectrometry. Current recommendations[7] are to use the latter term. Use of the term mass spectroscopy originated in the use of phosphor screens to detect ions.

- Mössbauer spectroscopy probes the properties of specific isotopic nuclei in different atomic environments by analyzing the resonant absorption of gamma-rays. See also Mössbauer effect.

- Neutron spin echo spectroscopy measures internal dynamics in proteins and other soft matter systems

- Photoacoustic spectroscopy measures the sound waves produced upon the absorption of radiation.

- Photoemission spectroscopy

- Photothermal spectroscopy measures heat evolved upon absorption of radiation.

- Pump-probe spectroscopy can use ultrafast laser pulses to measure reaction intermediates in the femtosecond timescale.

- Raman optical activity spectroscopy exploits Raman scattering and optical activity effects to reveal detailed information on chiral centers in molecules.

- Raman spectroscopy

- Saturated spectroscopy

- Scanning tunneling spectroscopy

- Spectrophotometry

- Time-resolved spectroscopy measures the decay rate(s) of excited states using various spectroscopic methods.

- Time-Stretch Spectroscopy[8][9]

- Thermal infrared spectroscopy measures thermal radiation emitted from materials and surfaces and is used to determine the type of bonds present in a sample as well as their lattice environment. The techniques are widely used by organic chemists, mineralogists, and planetary scientists.

- Ultraviolet photoelectron spectroscopy (UPS)

- Video spectroscopy

- Vibrational circular dichroism spectroscopy

- X-ray photoelectron spectroscopy (XPS)

13.5 Applications

- Estimate weathered wood exposure times using near infrared spectroscopy.[11]

- Cure monitoring of composites using Optical fibers

13.6 History

For main article see History of spectroscopy

The history of spectroscopy began with Isaac Newton's optics experiments (1666–1672). Newton applied the word "spectrum" to describe the rainbow of colors that combine to form white light and that are revealed when the white light is passed through a prism. During the early 1800s, Joseph von Fraunhofer made experimental advances with dispersive spectrometers that enabled spectroscopy to become a more precise and quantitative scientific technique. Since then, spectroscopy has played and continues to play a significant role in chemistry, physics and astronomy.

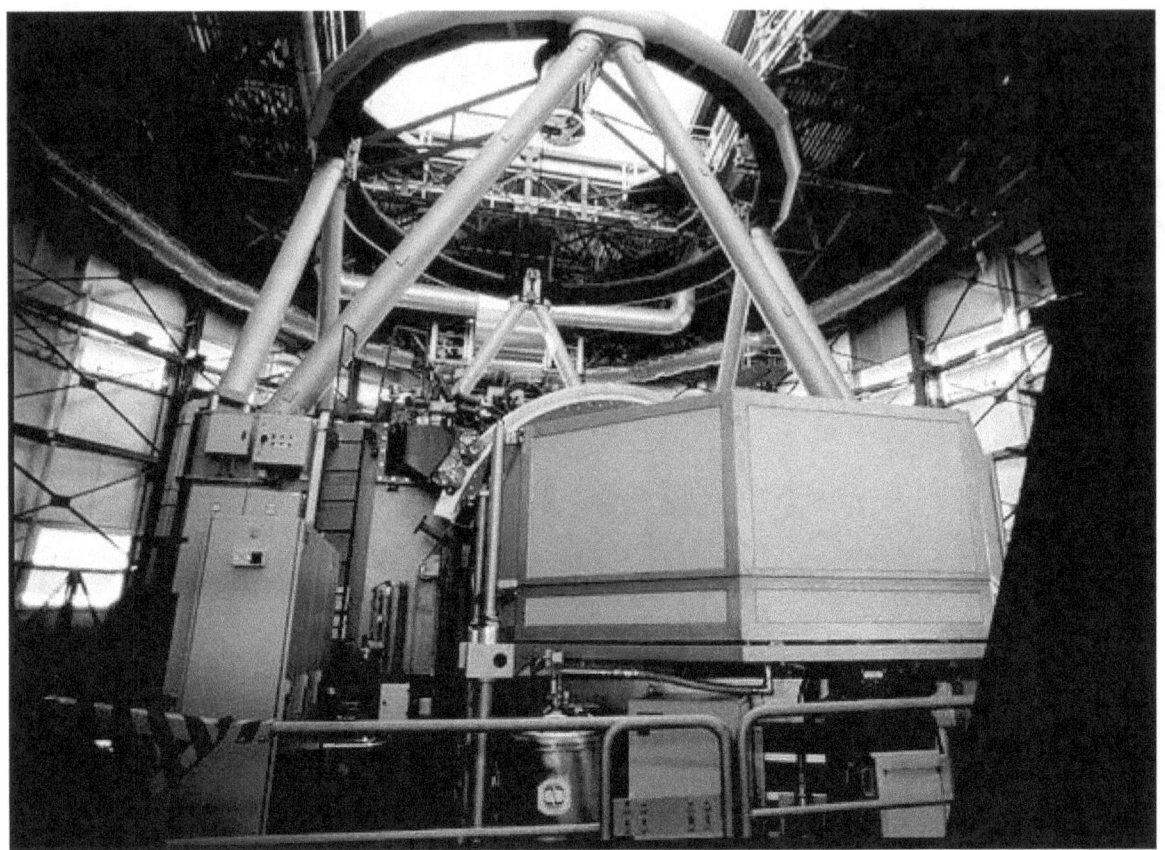

UVES is a high-resolution spectrograph on the Very Large Telescope.[10]

13.7 See also

- Astronomical spectroscopy
- Applied spectroscopy
- Biomedical spectroscopy
- History of spectroscopy
- List of spectroscopists
- Metamerism (color)
- Operando spectroscopy
- Scattering theory
- Spectral power distributions
- Spectroscopic notation

13.8 Notes

[1] Crouch, Stanley; Skoog, Douglas A. (2007). *Principles of instrumental analysis.* Australia: Thomson Brooks/Cole. ISBN 0-495-01201-7.

[2] Herrmann, R.; C. Onkelinx (1986). "Quantities and units in clinical chemistry: Nebulizer and flame properties in flame emission and absorption spectrometry (Recommendations 1986)". *Pure and Applied Chemistry* **58** (12): 1737–1742. doi:10.1351/pac.

[3] Mariani, Z.; Strong, K.; Wolff, M.; Rowe, P.; Walden, V.; Fogal, P. F.; Duck, T.; Lesins, G.; Turner, D. S.; Cox, C.; Eloranta, E.; Drummond, J. R.; Roy, C.; Turner, D. D.; Hudak, D.; Lindenmaier, I. A. (2012). "Infrared measurements in the Arctic using two Atmospheric Emitted Radiance Interferometers". *Atmos. Meas. Tech.* **5** (2): 329–344. Bibcode:2012AMT.....5..329M. doi:10.5194/amt-5-329-2012.

[4] Evans, C.L.; Xie, X.S. (2008). "Coherent Anti-Stokes Raman Scattering Microscopy: Chemical Imaging for Biology and Medicine".*Annual Review of Analytical Chemistry***1**: 883–909.Bibcode:2008ARAC....1..883E.doi:10.114.PMID20636101.

[5] W. Demtröder, *Laser Spectroscopy*, 3rd Ed. (Springer, 2003).

[6] F. J. Duarte (Ed.),*Tunable Laser Applications*, 2nd Ed. (CRC, 2009) Chapter 2.

[7] Murray, Kermit K.; Boyd, Robert K.; Eberlin, Marcos N.; Langley, G. John; Li, Liang; Naito, Yasuhide (2013). "Definitions of terms relating to mass spectrometry (IUPAC Recommendations 2013)". *Pure and Applied Chemistry* **85** (7): 1. doi:10.1351/PAC-REC-06-04-06. ISSN 0033-4545.

[8] Solli, D. R.; Chou, J.; Jalali, B. (2008). "Amplified wavelength–time transformation for real-time spectroscopy". *Nature Photonics* **2**: 48–51. Bibcode:2008NaPho...2...48S. doi:10.1038/nphoton.2007.253.

[9] Chou, Jason; Solli, Daniel R.; Jalali, Bahram (2008). "Real-time spectroscopy with subgigahertz resolution using amplified dispersive Fourier transformation". *Applied Physics Letters* **92** (11): 111102. arXiv:0803.1654. Bibcode:2008ApPhL..92k1102C. doi:10.1063/1.2896652.

[10] "Media advisory: Press Conference to Announce Major Result from Brazilian Astronomers". *ESO Announcement*. Retrieved 21 August 2013.

[11] Wang, Xiping; Wacker, James P. (2006). "Using NIR Spectroscopy to Predict Weathered Wood Exposure Times" (PDF). *WTCE 2006 – 9th world conference on timber engineering*.

13.9   References

- John M. Chalmers; Peter Griffiths, eds. (2006). *Handbook of Vibrational Spectroscopy*. New York: Wiley. doi:10.1002/0470027320. ISBN 0-471-98847-2. 5 Volume Set.

- Jerry Workman; Art Springsteen, eds. (1998). *Applied Spectroscopy*. Boston: Academic Press. ISBN 978-0-08-052749-9.

13.10   External links

- Spectroscopy links at DMOZ

- Amateur spectroscopy links at DMOZ

- NIST Atomic Spectroscopy Databases

- MIT Spectroscopy Lab's History of Spectroscopy

- Timeline of Spectroscopy

Chapter 14

Molecular geometry

Molecular geometry is the three-dimensional arrangement of the atoms that constitute a molecule. It determines several properties of a substance including its reactivity, polarity, phase of matter, color, magnetism, and biological activity.[1][2] The angles between bonds that an atom forms depend only weakly on the rest of molecule, i.e. they can be understood as approximately local and hence transferable properties.

14.1 Molecular geometry determination

The molecular geometry can be determined by various spectroscopic methods and diffraction methods. IR, microwave and Raman spectroscopy can give information about the molecule geometry from the details of the vibrational and rotational absorbance detected by these techniques. X-ray crystallography, neutron diffraction and electron diffraction can give molecular structure for crystalline solids based on the distance between nuclei and concentration of electron density. Gas electron diffraction can be used for small molecules in the gas phase. NMR and FRET methods can be used to determine complementary information including relative distances, [3][4][5] dihedral angles, [6][7] angles, and connectivity. Molecular geometries are best determined at low temperature because at higher temperatures the molecular structure is averaged over more accessible geometries (see next section). Larger molecules often exist in multiple stable geometries (conformational isomerism) that are close in energy on the potential energy surface. Geometries can also be computed by ab initio quantum chemistry methods to high accuracy. The molecular geometry can be different as a solid, in solution, and as a gas.

The position of each atom is determined by the nature of the chemical bonds by which it is connected to its neighboring atoms. The molecular geometry can be described by the positions of these atoms in space, evoking bond lengths of two joined atoms, bond angles of three connected atoms, and torsion angles (dihedral angles) of three consecutive bonds.

14.2 The influence of thermal excitation

Since the motions of the atoms in a molecule are determined by quantum mechanics, one must define "motion" in a quantum mechanical way. The overall (external) quantum mechanical motions translation and rotation hardly change the geometry of the molecule. (To some extent rotation influences the geometry via Coriolis forces and centrifugal distortion, but this is negligible for the present discussion.) In addition to translation and rotation, a third type of motion is molecular vibration, which corresponds to internal motions of the atoms such as bond stretching and bond angle variation. The molecular vibrations are harmonic (at least to good approximation), and the atoms oscillate about their equilibrium positions, even at the absolute zero of temperature. At absolute zero all atoms are in their vibrational ground state and show zero point quantum mechanical motion, so that the wavefunction of a single vibrational mode is not a sharp peak, but an exponential of finite width (the wavefunction for $n = 0$ depicted in the article on the quantum harmonic oscillator). At higher temperatures the vibrational modes may be thermally excited (in a classical interpretation one expresses this by stating that "the molecules will vibrate faster"), but they oscillate still around the recognizable geometry of the molecule.

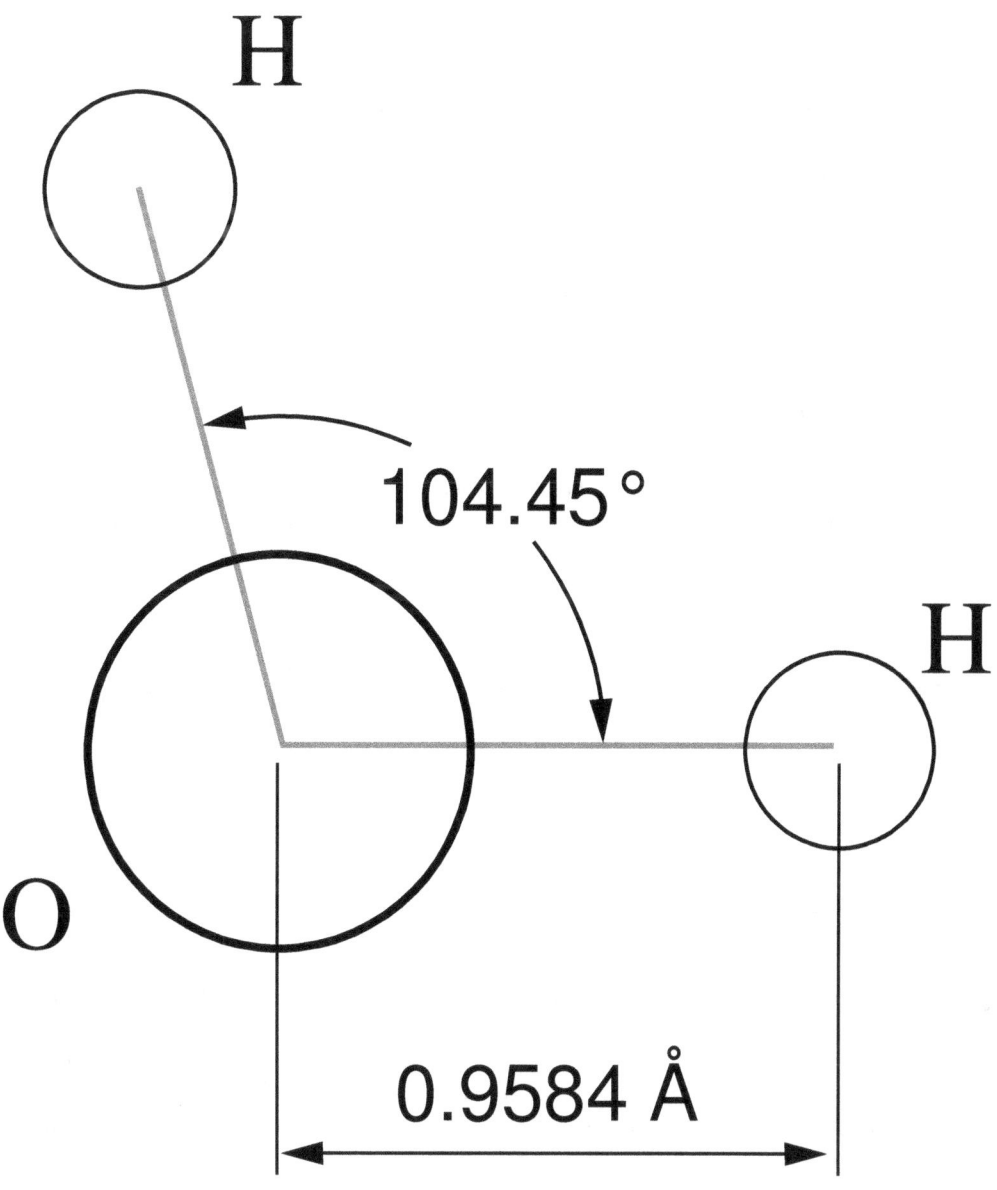

Geometry of the water molecule

To get a feeling for the probability that the vibration of molecule may be thermally excited, we inspect the Boltzmann factor $\beta \equiv \exp\left(-\frac{\Delta E}{kT}\right)$, where ΔE is the excitation energy of the vibrational mode, k the Boltzmann constant and T the absolute temperature. At 298 K (25 °C), typical values for the Boltzmann factor β are: $\beta = 0.089$ for $\Delta E = 500$ cm^{-1} ; $\beta = 0.008$ for $\Delta E = 1000$ cm^{-1} ; $\beta = 7 \times 10^{-4}$ for $\Delta E = 1500$ cm^{-1}. (The reciprocal centimeter is an energy unit that is commonly used in infrared spectroscopy; 1 cm^{-1} corresponds to 1.23984×10^{-4} eV). When an excitation energy is 500 cm^{-1}, then about 8.9 percent of the molecules are thermally excited at room temperature. To put this in perspective: the lowest excitation vibrational energy in water is the bending mode (about 1600 cm^{-1}). Thus, at room temperature less than 0.07 percent of all the molecules of a given amount of water will vibrate faster than at absolute zero.

As stated above, rotation hardly influences the molecular geometry. But, as a quantum mechanical motion, it is thermally excited at relatively (as compared to vibration) low temperatures. From a classical point of view it can be stated that at higher temperatures more molecules will rotate faster, which implies that they have higher angular velocity and angular momentum. In quantum mechanical language: more eigenstates of higher angular momentum become thermally populated with rising temperatures. Typical rotational excitation energies are on the order of a few cm^{-1}. The results of many spectroscopic experiments are broadened because they involve an averaging over rotational states. It is often difficult to extract geometries from spectra at high temperatures, because the number of rotational states probed in the experimental averaging increases with increasing temperature. Thus, many spectroscopic observations can only be expected to yield reliable molecular geometries at temperatures close to absolute zero, because at higher temperatures too many higher rotational states are thermally populated.

14.3 Bonding

Molecules, by definition, are most often held together with covalent bonds involving single, double, and/or triple bonds, where a "bond" is a shared pair of electrons (the other method of bonding between atoms is called ionic bonding and involves a positive cation and a negative anion).

Molecular geometries can be specified in terms of **bond lengths**, **bond angles** and **torsional angles**. The bond length is defined to be the average distance between the nuclei of two atoms bonded together in any given molecule. A bond angle is the angle formed between three atoms across at least two bonds. For four atoms bonded together in a chain, the torsional angle is the angle between the plane formed by the first three atoms and the plane formed by the last three atoms.

There exists a mathematical relationship among the bond angles for one central atom and four peripheral atoms (labeled 1 through 4) expressed by the following determinant. This constraint removes one degree of freedom from the choices of (originally) six free bond angles to leave only five choices of bond angles. (Note that the angles θ_{11}, θ_{22}, θ_{33}, and θ_{44} are always zero.)

$$0 = \begin{vmatrix} \cos\theta_{11} & \cos\theta_{12} & \cos\theta_{13} & \cos\theta_{14} \\ \cos\theta_{21} & \cos\theta_{22} & \cos\theta_{23} & \cos\theta_{24} \\ \cos\theta_{31} & \cos\theta_{32} & \cos\theta_{33} & \cos\theta_{34} \\ \cos\theta_{41} & \cos\theta_{42} & \cos\theta_{43} & \cos\theta_{44} \end{vmatrix}$$

Molecular geometry is determined by the quantum mechanical behavior of the electrons. Using the valence bond approximation this can be understood by the type of bonds between the atoms that make up the molecule. When atoms interact to form a chemical bond, the atomic orbitals of each atom are said to combine in a process called orbital hybridisation. The two most common types of bonds are sigma bonds (usually formed by hybrid orbitals) and pi bonds (formed by unhybridized p orbitals for atoms of main group elements). The geometry can also be understood by molecular orbital theory where the electrons are delocalised.

An understanding of the wavelike behavior of electrons in atoms and molecules is the subject of quantum chemistry.

14.4 Isomers

Isomers are types of molecules that share a chemical formula but have different geometries, resulting in very different properties:

- A **pure** substance is composed of only one type of isomer of a molecule (all have the same geometrical structure).

- Structural isomers have the same chemical formula but different physical arrangements, often forming alternate molecular geometries with very different properties. The atoms are not bonded (connected) together in the same orders.

    - Functional isomers are special kinds of structural isomers, where certain groups of atoms exhibit a special kind of behavior, such as an ether or an alcohol.

- Stereoisomers may have many similar physicochemical properties (melting point, boiling point) and at the same time very different biochemical activities. This is because they exhibit a handedness that is commonly found in living systems. One manifestation of this chirality or handedness is that they have the ability to rotate polarized light in different directions.

- Protein folding concerns the complex geometries and different isomers that proteins can take.

14.5   Types of molecular structure

A bond angle is the geometric angle between two adjacent bonds. Some common shapes of simple molecules include:

- **Linear:** In a linear model, atoms are connected in a straight line. The bond angles are set at $180°$. For example, carbon dioxide and nitric oxide have a linear molecular shape.

- **Trigonal planar:** Molecules with the trigonal planar shape are somewhat triangular and in one plane (flat). Consequently, the bond angles are set at $120°$. For example, boron trifluoride.

- **Bent:** Bent or angular molecules have a non-linear shape. For example, water (H_2O), which has an angle of about $105°$. A water molecule has two pairs of bonded electrons and two unshared lone pairs.

- **Tetrahedral:** *Tetra-* signifies four, and *-hedral* relates to a face of a solid, so "tetrahedral" literally means "having four faces". This shape is found when there are four bonds all on one central atom, with no extra unshared electron pairs. In accordance with the VSEPR (valence-shell electron pair repulsion theory), the bond angles between the electron bonds are $\arccos(-1/3) = 109.47°$. For example, methane (CH_4) is a tetrahedral molecule.

- **Octahedral:** *Octa-* signifies eight, and *-hedral* relates to a face of a solid, so "octahedral" means "having eight faces". The bond angle is 90 degrees. For example, sulfur hexafluoride (SF_6) is an octahedral molecule.

- **Trigonal pyramidal:** A trigonal pyramidal molecule has a pyramid-like shape with a triangular base. Unlike the linear and trigonal planar shapes but similar to the tetrahedral orientation, pyramidal shapes require three dimensions in order to fully separate the electrons. Here, there are only three pairs of bonded electrons, leaving one unshared lone pair. Lone pair – bond pair repulsions change the bond angle from the tetrahedral angle to a slightly lower value.[8] For example, ammonia (NH_3).

14.5.1   VSEPR table

Main article: VSEPR theory § AXE method

The bond angles in the table below are ideal angles from the simple VSEPR theory, followed by the actual angle for the example given in the following column where this differs. For many cases, such as trigonal pyramidal and bent, the actual angle for the example differs from the ideal angle, but all examples differ by different amounts. For example, the angle in H_2S ($92°$) differs from the tetrahedral angle by much more than the angle for H_2O ($104.48°$) does.

14.6   3D representations

- **Line** or **stick** – atomic nuclei are not represented, just the bonds as sticks or lines. As in 2D molecular structures of this type, atoms are implied at each vertex.

:{| class=wikitable |- | |

|

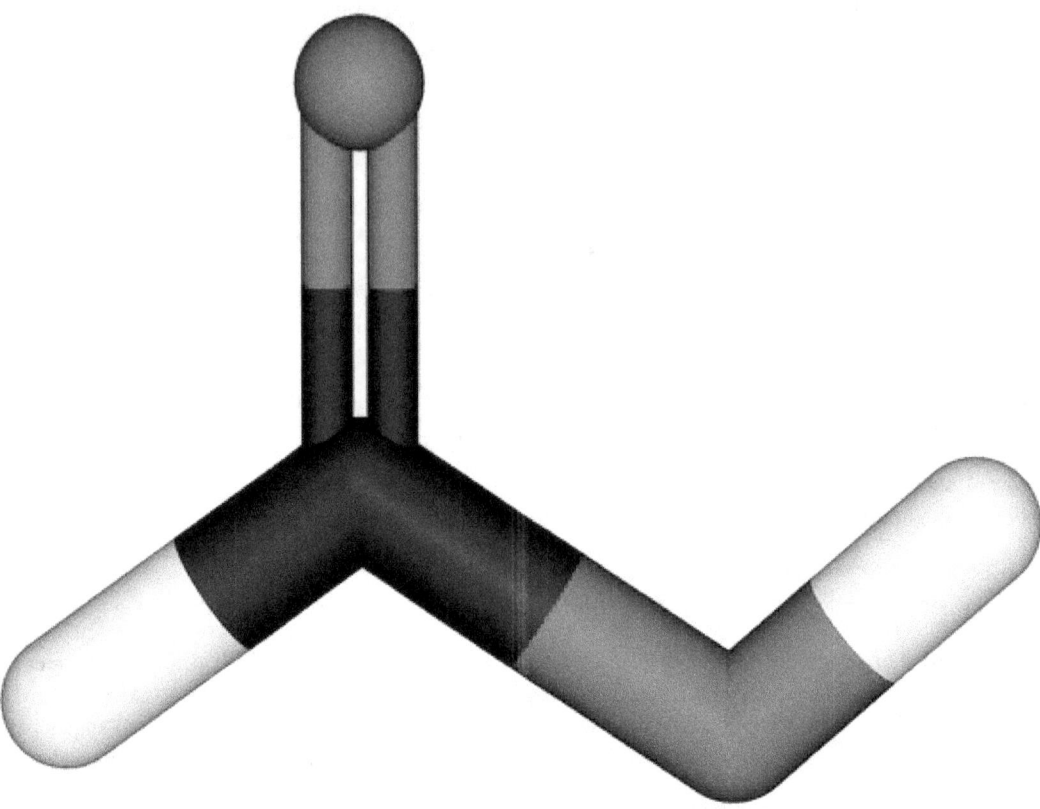

|
|}

- **Electron density plot** – shows the electron density determined either crystallographically or using quantum mechanics rather than distinct atoms or bonds.

- **Ball and stick** – atomic nuclei are represented by spheres (balls) and the bonds as sticks.

:{| class=wikitable |- | |

|

|

|}

- Spacefilling models or CPK models (also an atomic coloring scheme in representations) – the molecule is represented by overlapping spheres representing the atoms.

:{| class=wikitable |- | |

|

|

I}

- **Cartoon** – a representation used for proteins where loops, beta sheets, alpha helices are represented diagrammatically and no atoms or bonds are represented explicitly just the protein backbone as a smooth pipe.

The greater the amount of lone pairs contained in a molecule the smaller the angles between the atoms of that molecule. The VSEPR theory predicts that lone pairs repel each other, thus pushing the different atoms away from them.

14.7   See also

- Molecular graphics

- Molecular modelling

- Molecular mechanics

- Molecule editor

- Molecular design software

- Quantum chemistry

- Polyhedral skeletal electron pair theory

- Topology (chemistry)

- Jemmis mno rules

14.8   References

[1] McMurry, John E. (1992), *Organic Chemistry* (3rd ed.), Belmont: Wadsworth, ISBN 0-534-16218-5

[2] Cotton, F. Albert; Wilkinson, Geoffrey; Murillo, Carlos A.; Bochmann, Manfred (1999), *Advanced Inorganic Chemistry* (6th ed.), New York: Wiley-Interscience, ISBN 0-471-19957-5

[3] FRET description

[4] Hillisch, A; Lorenz, M; Diekmann, S (2001). "Recent advances in FRET: distance determination in protein–DNA complexes". *Current Opinion in Structural Biology* **11** (2): 201–207. doi:10.1016/S0959-440X(00)00190-1. PMID 11297928.

[5] FRET imaging introduction

[6] obtaining dihedral angles from ^{3}J coupling constants

[7] Another Javascript-like NMR coupling constant to dihedral

[8] Miessler G.L. and Tarr D.A. *Inorganic Chemistry* (2nd ed., Prentice-Hall 1999), pp.57-58

[9] Hoy, AR; Bunker, PR (1979). "A precise solution of the rotation bending Schrödinger equation for a triatomic molecule with application to the water molecule". *Journal of Molecular Spectroscopy* **74**: 1–8. doi:10.1016/0022-2852(79)90019-5.

[10] http://cccbdb.nist.gov/expangle2.asp?descript=aHOH&all=0

14.9   External links

- Molecular Geometry & Polarity Tutorial 3D visualization of molecules to determine polarity.

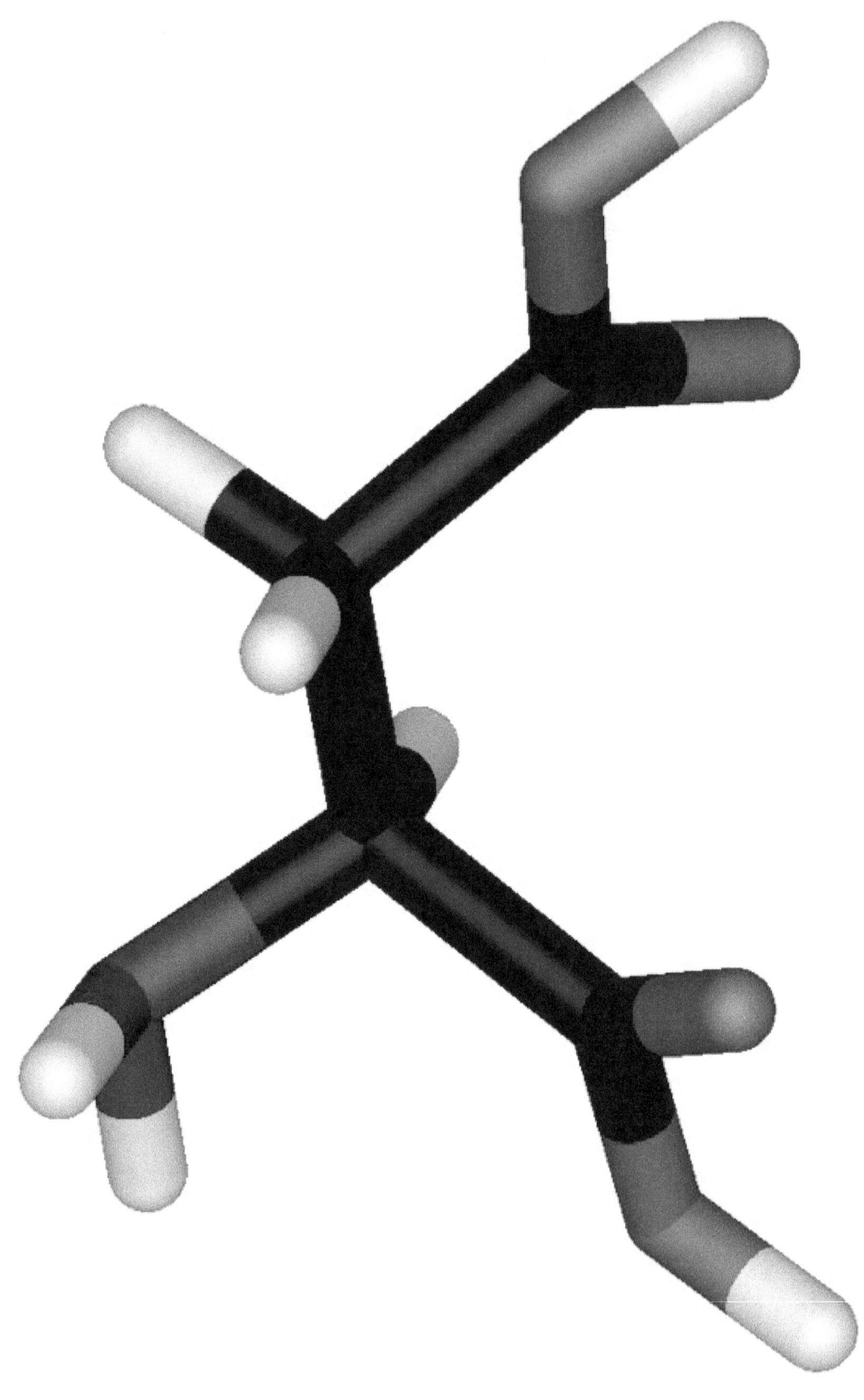

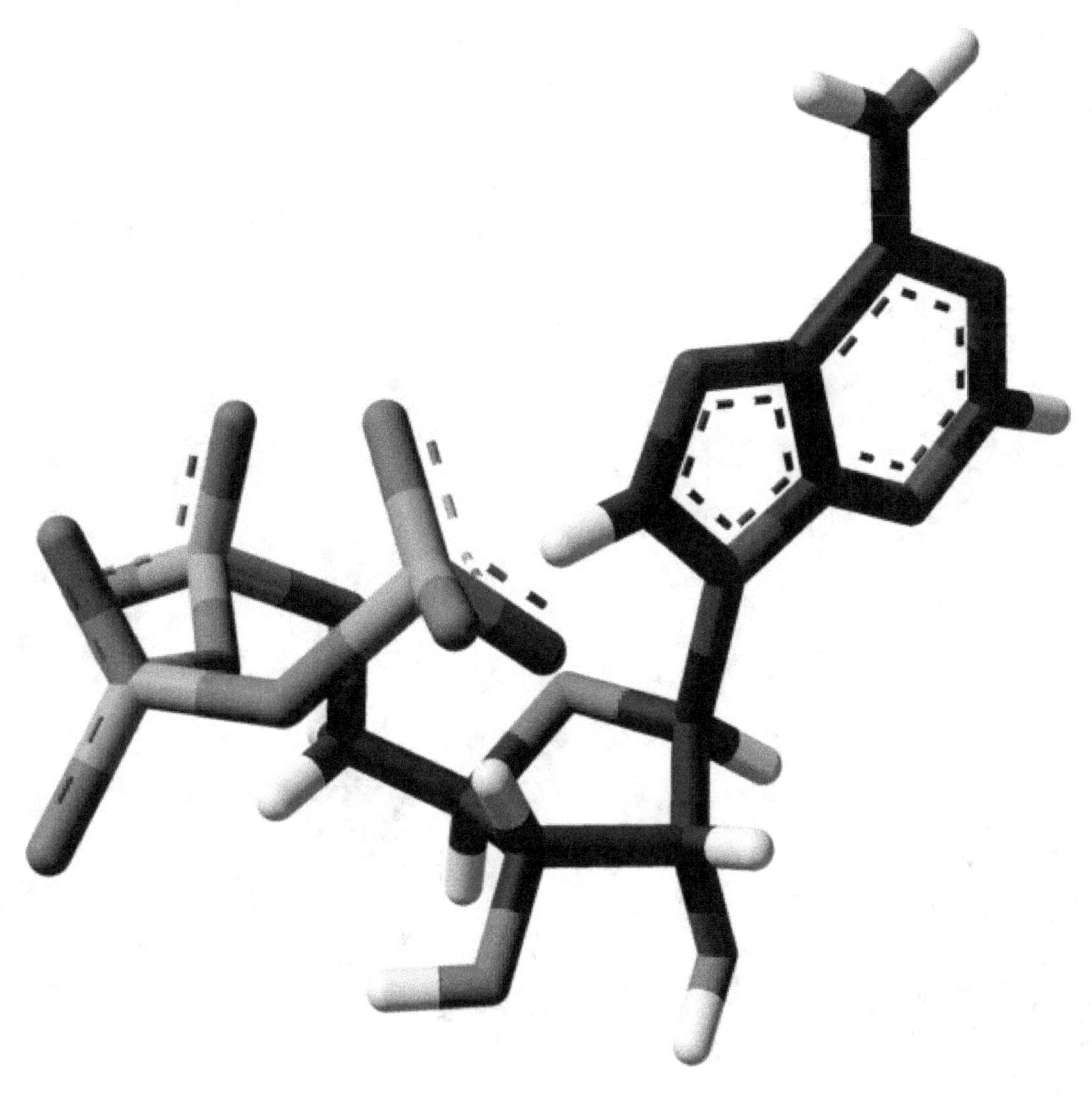

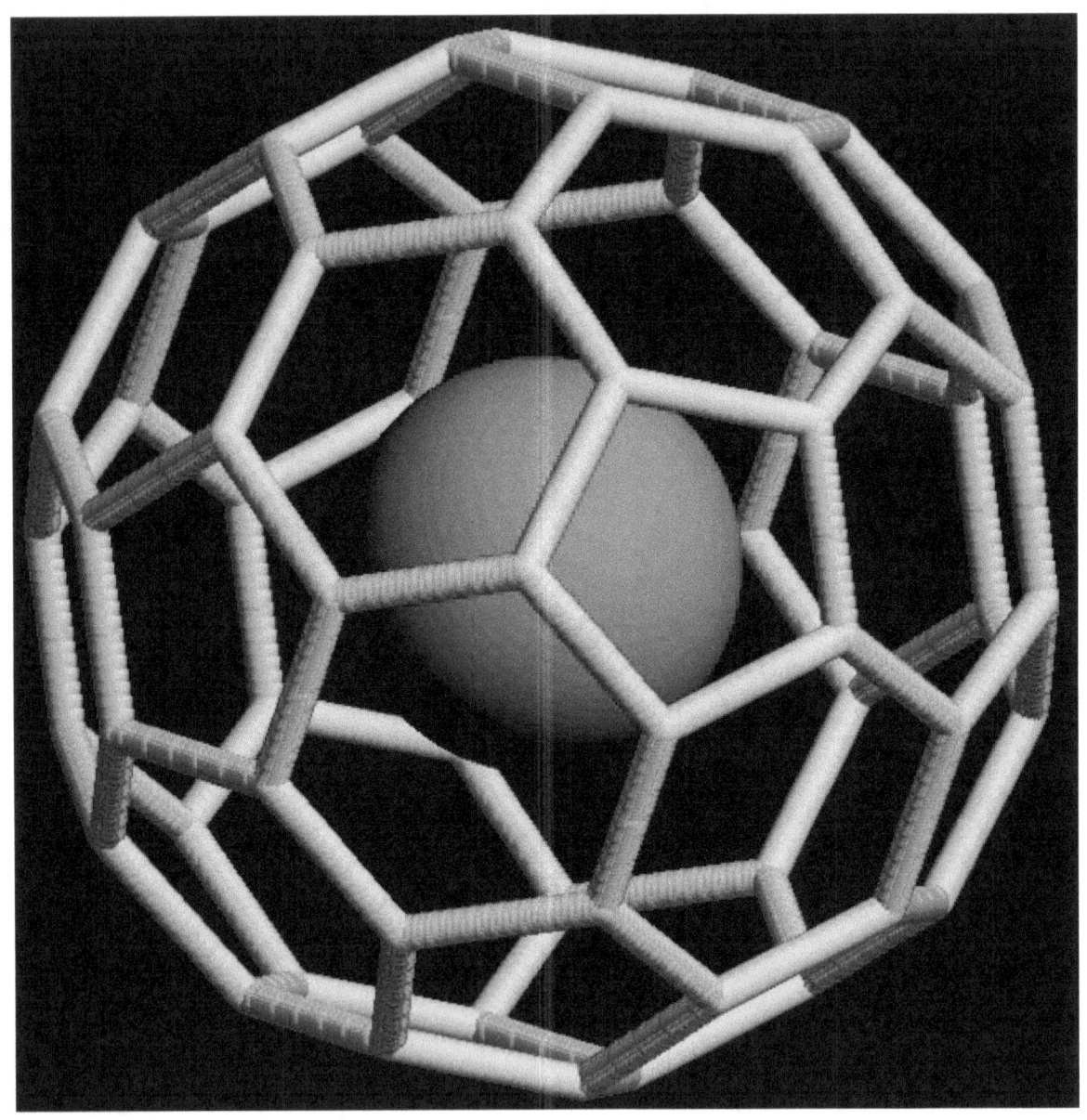

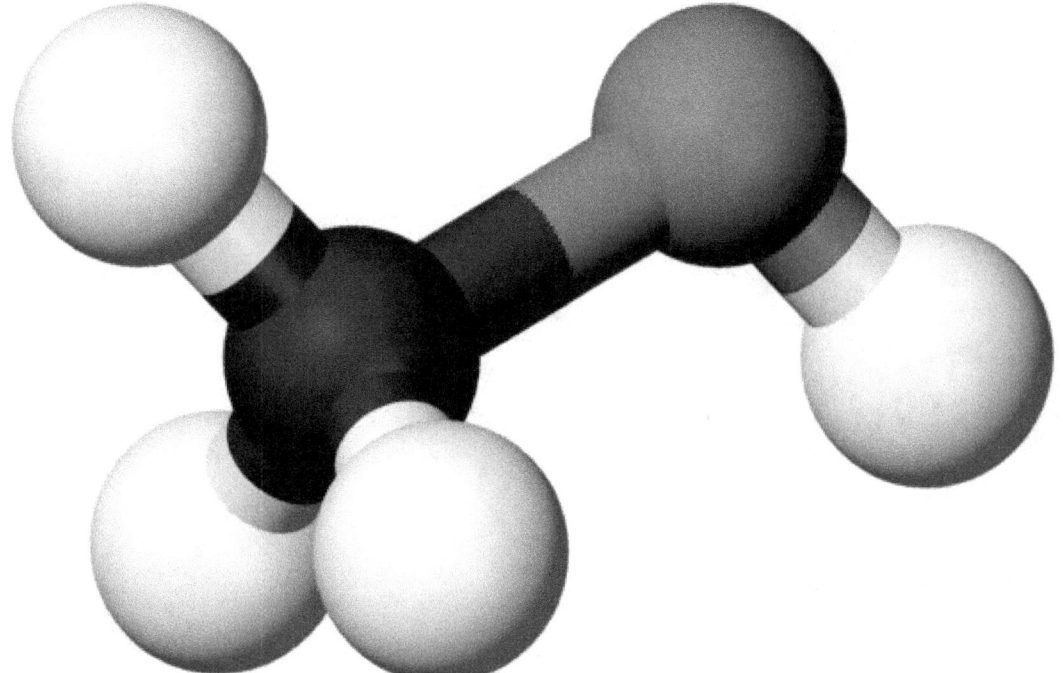

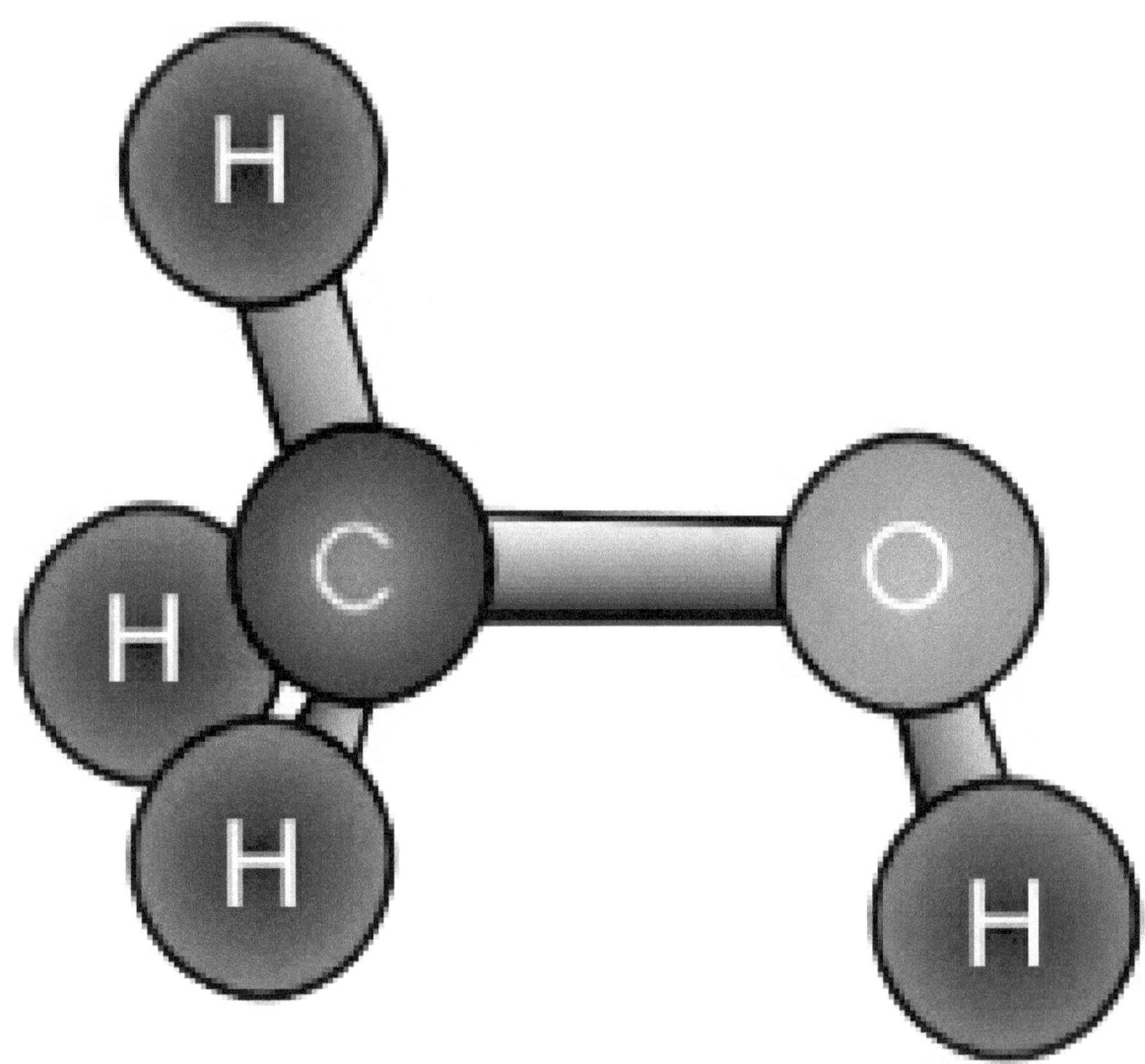

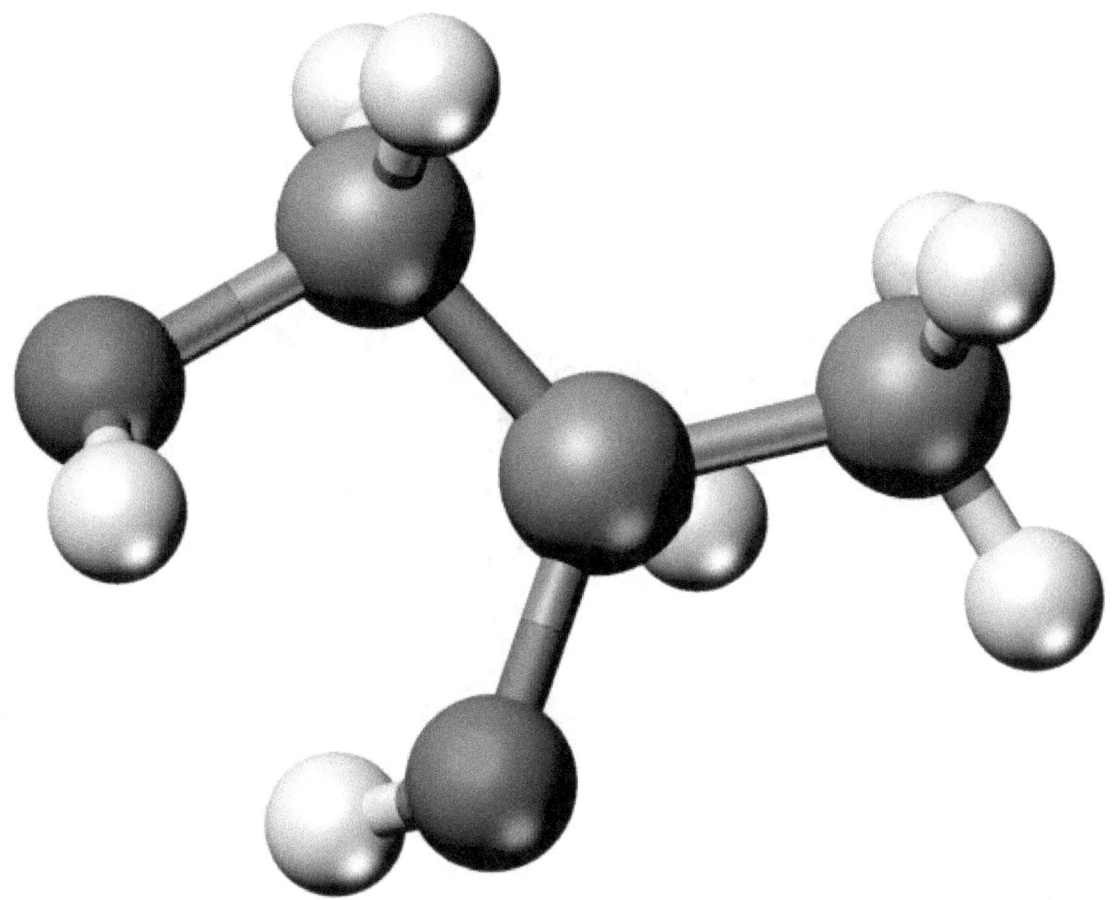

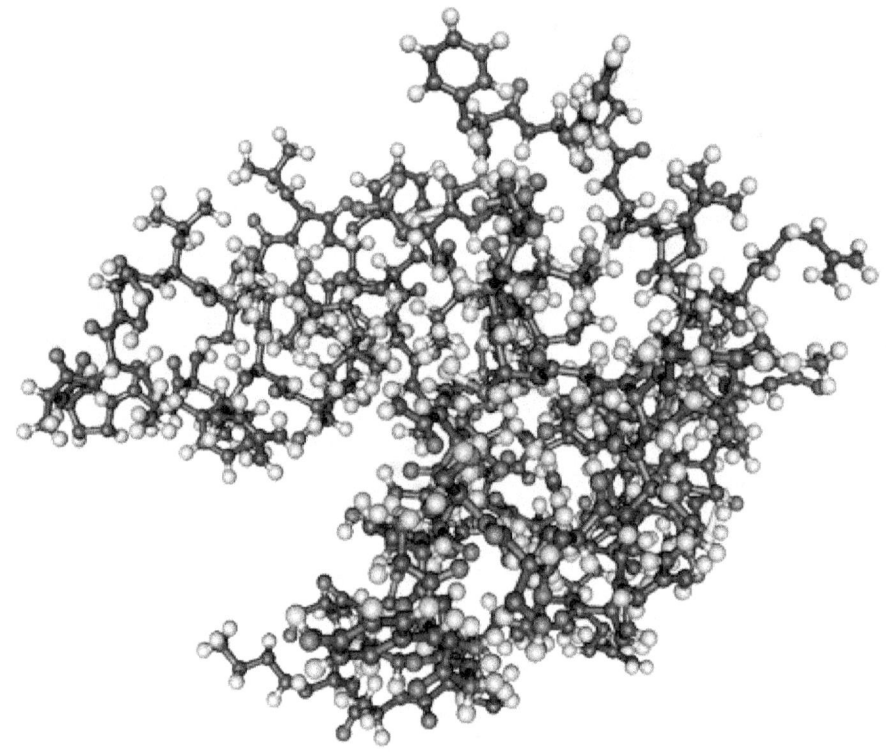

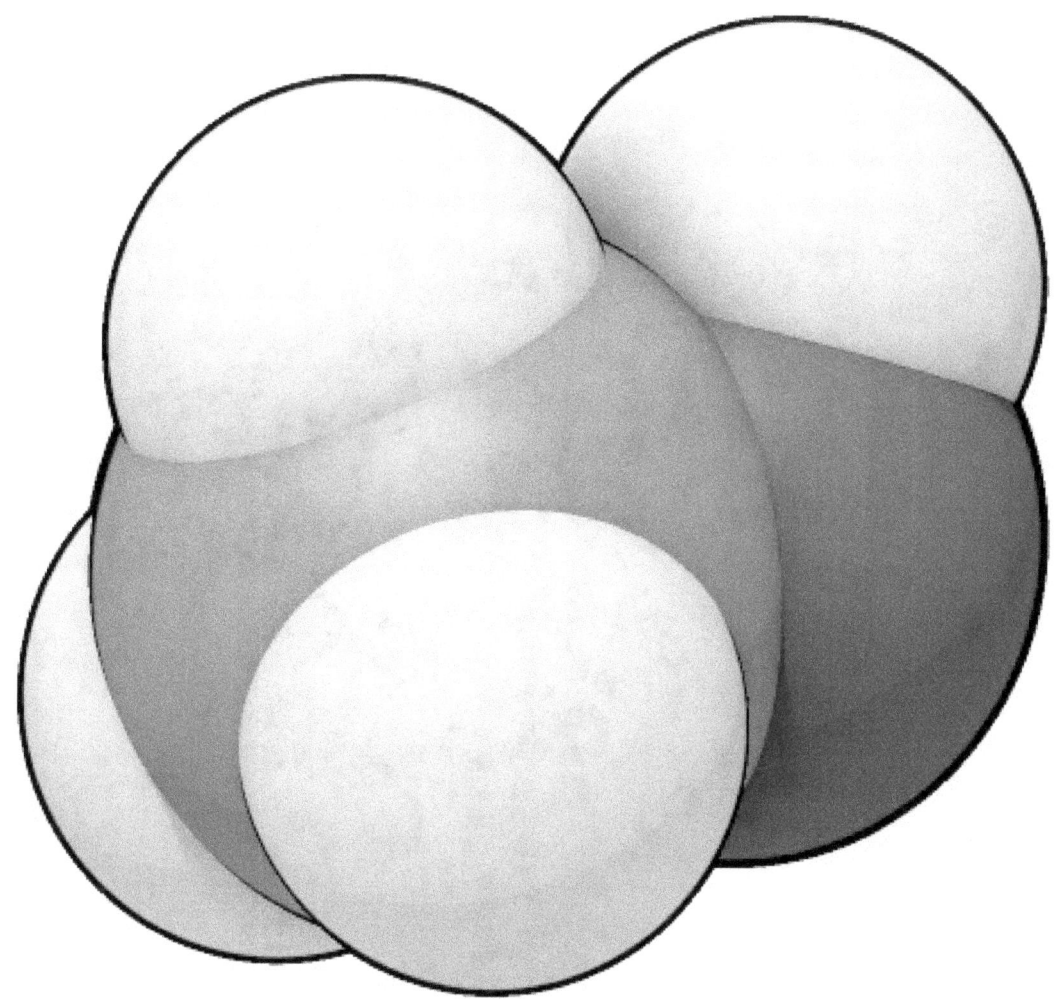

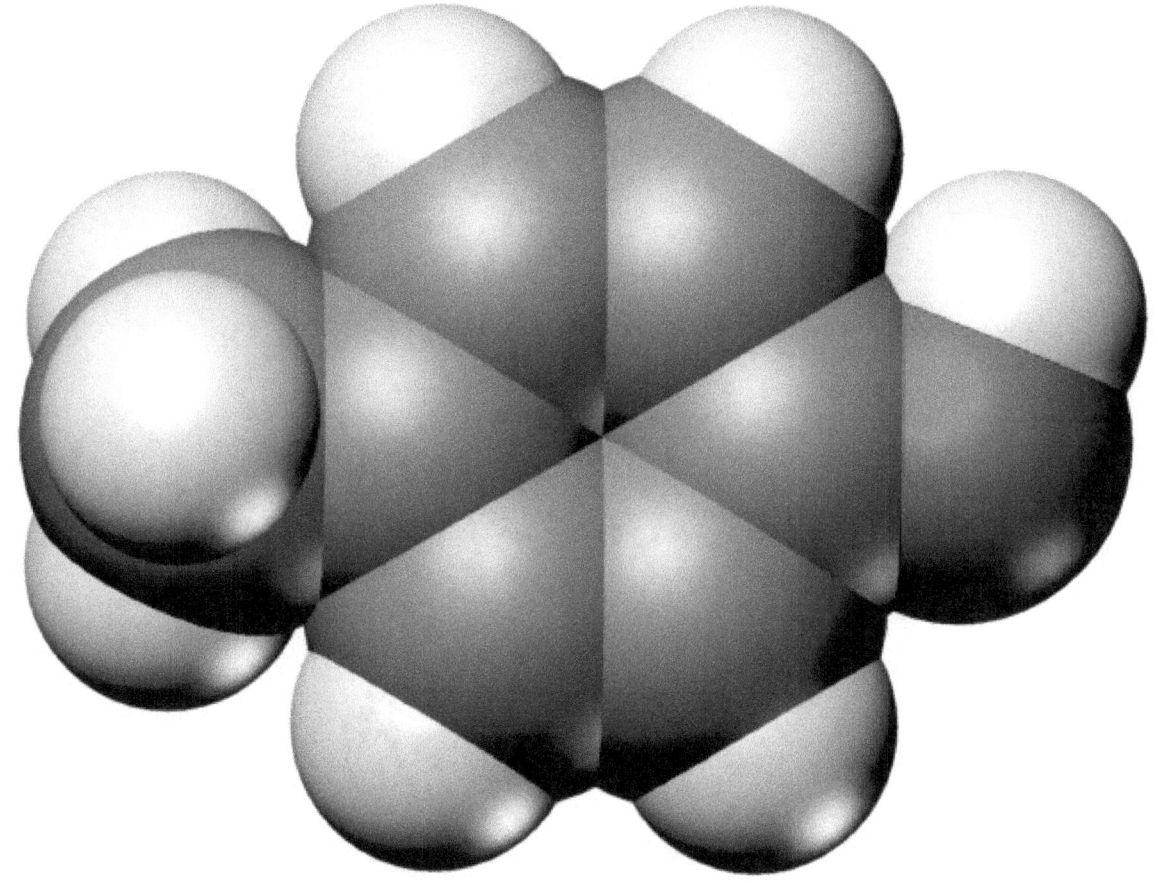

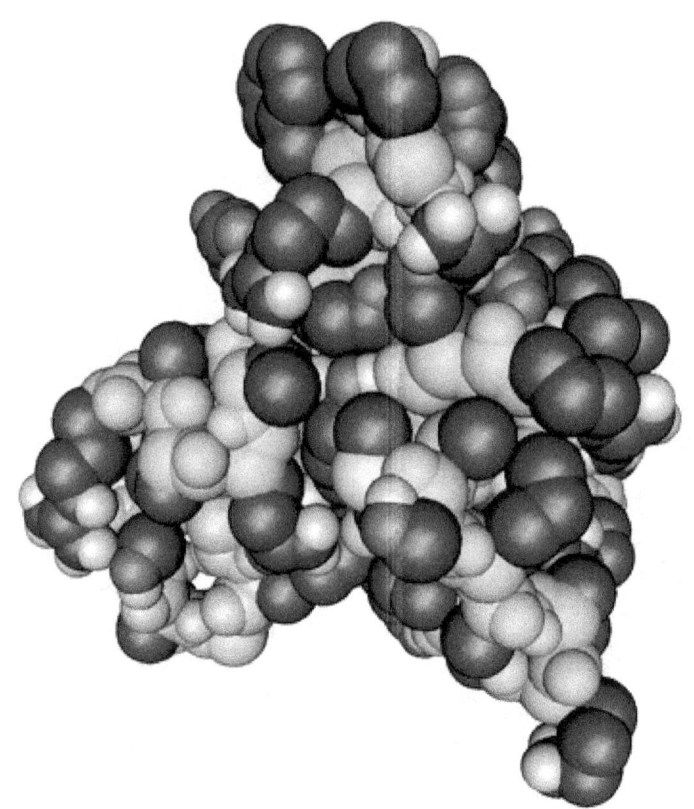

Chapter 15

Periodic systems of small molecules

Periodic systems of molecules are charts of molecules similar to the periodic table of the elements. Construction of such charts was initiated in the early 20th century and is still ongoing.

It is commonly believed that the periodic law, represented by the periodic chart, is echoed in the behavior of molecules, at least small molecules. For instance, if one replaces any one of the atoms in a triatomic molecule with a rare gas atom, there will be a drastic change in the molecule's properties. Several goals could be accomplished by constructing an explicit representation of this periodic law as manifested in molecules: (1) a classification scheme for the vast number of molecules that exist, starting with small ones having just a few atoms, for use as a teaching aid and tool for archiving data, (2) forecasting data for molecular properties based on the classification scheme, and (3) a sort of unity with the periodic chart and the periodic system of fundamental particles.[1]

15.1   Physical periodic systems of molecules

Periodic systems (or charts or tables) of molecules are the subjects of two reviews.[2][3] The systems of diatomic molecules include those of (1) H. D. W. Clark,[4][5] and (2) F.-A. Kong,[6][7] which somewhat resemble the atomic chart. The system of R. Hefferlin *et al.*[8][9] was developed from (3) a three-dimensional to (4) a four-dimensional system Kronecker product of the element chart with itself.

A totally different kind of periodic system is (5) that of G. V. Zhuvikin,[11][12] which is based on group dynamics. In all but the first of these cases, other researchers provided invaluable contributions and some of them are co-authors. The architectures of these systems have been adjusted by Kong[7] and Hefferlin [13] to include ionized species, and expanded by Kong,[7] Hefferlin,[9] and Zhuvikin and Hefferlin[12] to the space of triatomic molecules. These architectures are mathematically related to the chart of the elements. They were first called "physical" periodic systems.[2]

15.2   Chemical periodic systems of molecules

Other investigators have focused on building structures that address specific kinds of molecules such as alkanes(Morozov); benzenoids (Dias) ;[15][16]functional groups containing fluorine,oxygen, nitrogen and sulfur(Haas);[17][18]or a combination of core charge, number of shells, redox potentials, and acid-base tendencies (Gorski).[19][20]These structures are not restricted to molecules with a given number of atoms and they bear little resemblance to the element chart; they are called "chemical" systems. Chemical systems do not start with the element chart, but instead start with, for example , formulae numerations(Dias), the hydrogen-displacement principle (Haas), reduced potential curves(Jenz),[21]a set of molecular descriptors (Gorski), and similar strategies.

15.3 Hyperperiodicity

E. V. Babaev[22] has erected a hyperperiodic system which in principle includes all of the systems described above except those of Dias, Gorski, and Jenz.

15.4 Bases of the element chart and periodic systems of molecules

The periodic chart of the elements, like a small stool, is supported by three legs: (a) the Bohr–Sommerfeld "solar system" atomic model (with electron spin and the Madelung principle), which provides the magic-number elements that end each row of the table and gives the number of elements in each row, (b) solutions to the Schrödinger equation, which provide the same information, and (c) data provided by experiment, by the solar system model, and by solutions to the Schroedinger equation. The Bohr–Sommerfeld model should not be ignored: it gave explanations for the wealth of spectroscopic data that were already in existence before the advent of wave mechanics.

Each of the molecular systems listed above, and those not cited, is also supported by three legs: (a) physical and chemical data arranged in graphical or tabular patterns (which, for physical periodic systems at least, echo the appearance of the element chart), (b) group dynamic, valence-bond, molecular-orbital, and other fundamental theories, and (c) summing of atomic period and group numbers (Kong), the Kronecker product and exploitation of higher dimensions (Hefferlin), formula enumerations (Dias), the hydrogen-displacement principle (Haas), reduced potential curves (Jenz), and similar strategies.

A chronological list of the contributions to this field[3] contains almost thirty entries dated 1862, 1907, 1929, 1935, and 1936; then, after a pause, a higher level of activity beginning with the 100th anniversary of Mendeleev's publication of his element chart, 1969. Many publications on periodic systems of molecules include some predictions of molecular properties, but starting at the turn of the Century there have been serious attempts to use periodic systems for the prediction of progressively more precise data for various numbers of molecules. Among these attempts are those of Kong,[7] and Hefferlin[23][24]

15.5 A collapsed-coordinate system for triatomic molecules

The collapsed-coordinate system has three independent variables instead of the six demanded by the Kronecker-product system. The reduction of independent variables makes use of three properties of gas-phase, ground-state, triatomic molecules. (1) In general, whatever the total number of constituent atomic valence electrons, data for isoelectronic molecules tend to be more similar than for adjacent molecules that have more or fewer valence electrons; for triatomic molecules, the electron count is the sum of the atomic group numbers (the sum of the column numbers 1 to 8 in the p-block of the periodic chart of the elements, $C1+C2+C3$). (2) Linear/bent triatomic molecules appear to be slightly more stable, other parameters being equal, if carbon is the central atom. (3) Most physical properties of diatomic molecules (especially spectroscopic constants) are closely monotonic with respect to the product of the two atomic period (or row) numbers, R1 and R2; for triatomic molecules, the monotonicity is close with respect to $R1R2+R2R3$ (which reduces to $R1R2$ for diatomic molecules). Therefore, the coordinates x, y, and z of the collapsed-coordinate system are $C1+C2+C3$, C2, and $R1R2+R2R3$. Multiple-regression predictions of four property values for molecules with tabulated data agree very well with the tabulated data (the error measures of the predictions include the tabulated data in all but a few cases).[25]

15.6 References

[1] Chung, D.-Y. (2000). "The Periodic Table of Elementary Particles". arXiv:physics/0003023.

[2] Hefferlin, R. and Burdick, G.W. 1994. Fizicheskie i khimicheskie periodicheskie sistemy Molekul, Zhurnal Obshchei Xhimii, vol. 64, pp. 1870–1885. English translation: "Periodic Systems of Molecules: Physical and Chemical". *Russ. J. Gen. Chem.* **64**: 1659–1674.

[3] Hefferlin, R. 2006. The Periodic Systems of Molecules pp. 221 ff, in Baird, D., Scerri, E., and McIntyre, L. (Eds.) "The Philosophy of Chemistry, Synthesis of a New Discipline," Springer, Dordrecht ISBN 1-4020-3256-0.

[4] Clark, C. H. D. (1935). "The periodic Groups of Non-Hydride Di-Atoms".*Trans. Faraday Soc* **31**: 1017–1036.doi:10.1039/tf9.

[5] Clark, C. H. D (1940). "Systematics of Band-Spectral Constants. Part V. Interrelations of Dissociation Energy and Equilibrium Internuclear Distance of Di-Atoms in Ground States". *Trans. Faraday Soc.* **36**: 370–376.

[6] Kong, F (1982). "The Periodicity of Diatomic Molecules".*J. Mol. Struct* **90**: 17–28.Bibcode:1982JMoSt..90...17K.doi:10.1016-2860(82)90199-5.

[7] Kong, F. and Wu, W. 2010. Periodicity of Diatomic and Triatomic Molecules, Conference Proceedings of the 2010 Workshop on Mathematical Chemistry of the Americas.

[8] Hefferlin, R., Campbell, D. Gimbel, H. Kuhlman, and T. Cayton (1979). "The periodic table of diatomic molecules—I an algorithm for retrieval and prediction of spectrophysical properties". *Quant. Spectrosc. Radiat. Transfer* **21** (4): 315–336. Bibcode:1979JQSRT..21..315H. doi:10.1016/0022-4073(79)90063-3.

[9] Hefferlin, R (2008). "Kronecker-Product Periodic Systems of Small Gas-Phase Molecules and the Search for Order in Atomic Ensembles of Any Phase". *Comb. Chem. High Through. Screen* **11**: 690–706.

[10] Gary W. Burdick and Ray Hefferlin, "Chapter 7. Data Location in a Four-Dimensional Periodic System of Diatomic Molecules", in Mihai V Putz, Ed., Chemical Information and Computational Challenges in the 21st Century, NOVA, 2011, ISBN 978-1-61209-712-1

[11] Zhuvikin, G.V. and R. Hefferlin (1983). "Periodicheskaya Sistema Dvukhatomnykh Molekul: Teoretiko-gruppovoi Podkhod, Vestnik Leningradskovo Universiteta" (16). pp. 10–16.

[12] Carlson, C.M., Cavanaugh, R.J, Hefferlin, R.A, and of Zhuvikin, G.V. (1996). "Periodic Systems of Molecular States from the Boson Group Dynamics of SO(3)xSU(2)s". *Chem. Inf. Comp. Sci* **36**: 396–398. doi:10.1021/ci9500748.

[13] Hefferlin, R. et al. (1984). "Periodic Systems of N-atom Molecules". *J. Quant. Spectrosc. Radiat. Transfer* **32** (4): 257–268. Bibcode:1984JQSRT..32..257H. doi:10.1016/0022-4073(84)90098-0.

[14] Morozov, N. 1907. Stroeniya Veshchestva, I. D. Sytina Publication, Moscow.

[15] Dias, J.R. (1982). "A periodic Table of Polycyclic Aromatic Hydrocarbons. Isomer Enumeration of Fused Polycyclic Aromatic Hydrocarbons". *Chem. Inf. Comput. Sci.* **22**: 15–22. doi:10.1021/ci00033a004.

[16] Dias, J. R. (1994). "Benzenoids to Fullerines and the Circumscribing and Leapfrog Algorithms". *New J. Chem.* **18**: 667–673.

[17] Haas, A. (1982). "A new classification principle: the periodic system of functional groups". *Chemicker-Zeitung* **106**: 239–248.

[18] Haas, A. (1988). "Das Elementverscheibungsprinzip und siene Bedeutung fur die Chemie der p-Block Elemente". *Kontakte (Darmstadt)* **3**: 3–11.

[19] Gorski, A (1971). "Morphological Classification of Simple Species. Part I. Fundamental Components of Chemical Structure". *Roczniki Chemii* **45**: 1981–1989.

[20] Gorski, A (1973). "Morphological Classification of Simple Species. Part V. Evaluation of Structural Parameters of Species". *Roczniki Chemii* **47**: 211–216.

[21] Jenz, F (1996). "The Reduced Potential Curve (RPC) Method and its Applications". *Int. Rev. Phys. Chem.* **15** (2): 467–523. Bibcode:1996IRPC...15..467J. doi:10.1080/01442359609353191.

[22] Babaev, E.V. and R. Hefferlin 1996. The Concepts of Periodicity and Hyper- periodicity: from Atoms to Molecules, in Rouvray, D.H. and Kirby, E.C., "Concepts in Chemistry," Research Studies Press Limited, Taunton, Somerset, England.

[23] Hefferlin, R. (2010). "Vibration Frequencies using Least squares and Neural Networks for 50 new s and p Electron Diatomics". *Quant. Spectr. Radiat. Transf.* **111**: 71–77. Bibcode:2010JQSRT.111...71H. doi:10.1016/j.jqsrt.2009.08.004.

[24] Hefferlin, R. (2010). "Internuclear Separations using Least squares and Neural Networks for 46 new s and p Electron Diatomics".

[25] Carlson, C., Gilkeson, J., Linderman, K., LeBlanc, S. Hefferlin, R., and Davis, B (1997). "Estimation of Properties of Triatomic Molecules from Tabulated Data Using Least-Squares Fitting". *Croatica Chemica Acta* **70**: 479–508.

Chapter 16

Macromolecule

For the scientific journal, see Macromolecules (journal).

A **macromolecule** is a very large molecule commonly created by polymerization of smaller subunits (monomers).

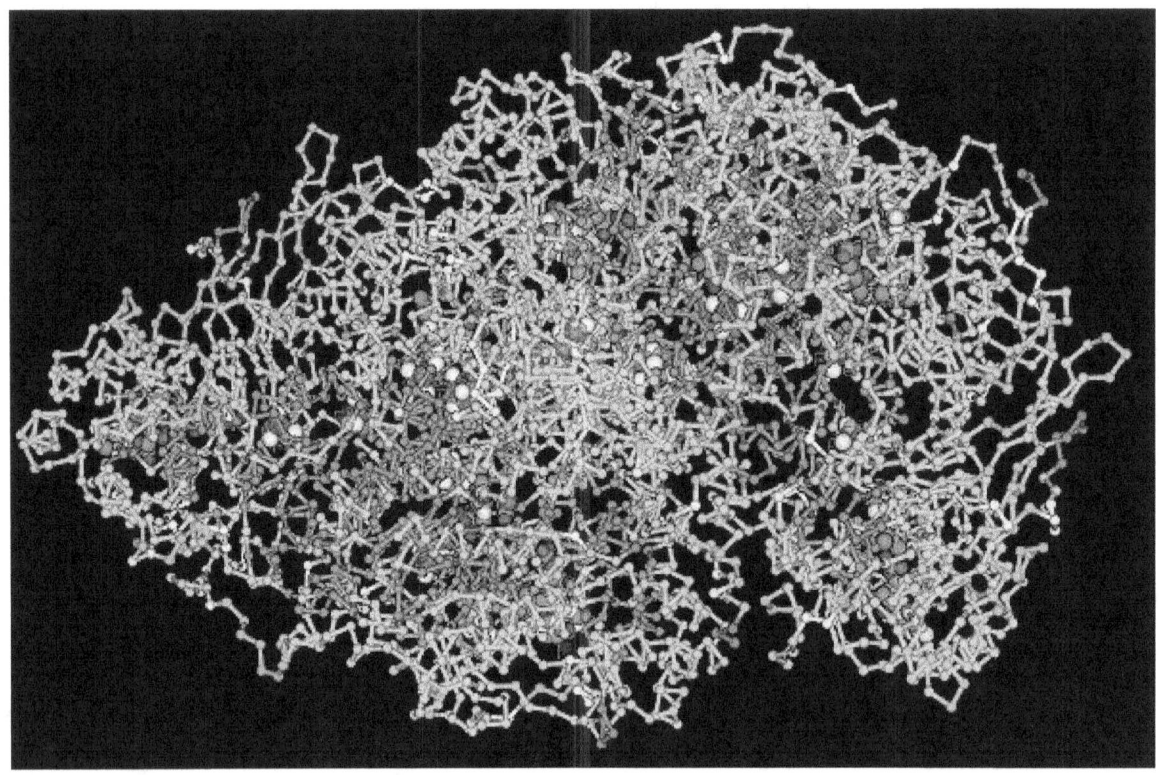

Chemical structure of a polypeptide macromolecule

They are typically composed of thousands or more atoms. The most common macromolecules in biochemistry are biopolymers (nucleic acids, proteins, carbohydrates and polyphenols) and large non-polymeric molecules (such as lipids and macrocycles).[1] Synthetic macromolecules include common plastics and synthetic fibres as well as experimental materials such as carbon nanotubes.[2][3]

16.1 Definition

IUPAC definition

Macromolecule
Polymer molecule

A molecule of high relative molecular mass, the structure of which essentially comprises the multiple repetition of units derived, actually or conceptually, from molecules of low relative molecular mass.

Notes

1. In many cases, especially for synthetic polymers, a molecule can be regarded as having a high relative molecular mass if the addition or removal of one or a few of the units has a negligible effect on the molecular properties. This statement fails in the case of certain macromolecules for which the properties may be critically dependent on fine details of the molecular structure.
2. If a part or the whole of the molecule fits this definition, it may be described as either *macromolecular* or *polymeric*, or by *polymer* used adjectivally.[4]

The term *macromolecule* (*macro* + *molecule*) was coined by Nobel laureate Hermann Staudinger in the 1920s, although his first relevant publication on this field only mentions *high molecular compounds* (in excess of 1,000 atoms).[5] At that time the phrase *polymer*, as introduced by Berzelius in 1833, had a different meaning from that of today: it simply was another form of isomerism for example with benzene and acetylene and had little to do with size.[6]

Usage of the term to describe large molecules varies among the disciplines. For example, while biology refers to macromolecules as the four large molecules comprising living things, in chemistry, the term may refer to aggregates of two or more molecules held together by intermolecular forces rather than covalent bonds but which do not readily dissociate.[7]

According to the standard IUPAC definition, the term *macromolecule* as used in polymer science refers only to a single molecule. For example,a single polymeric molecule is appropriately described as a "macromolecule" or "polymer molecule" rather than a "polymer", which suggests a substance composed of macromolecules.[8]

Because of their size, macromolecules are not conveniently described in terms of stoichiometry alone. The structure of simple macromolecules, such as homopolymers, may be described in terms of the individual monomer subunit and total molecular mass. Complicated biomacromolecules, on the other hand, require multi-faceted structural description such as the hierarchy of structures used to describe proteins. In British English, the word "macromolecule" tends to be called **"high polymer"**. [9]

16.2 Properties

Macromolecules often have unusual physical properties that do not occur for smaller molecules.

For example, DNA in a solution can be broken simply by sucking the solution through an ordinary straw because the physical forces on the molecule can overcome the strength of its covalent bonds. The 1964 edition of Linus Pauling's *College Chemistry* asserted that DNA in nature is never longer than about 5,000 base pairs.[10] This error arose because biochemists were inadvertently breaking their samples into fragments. In fact, the DNA of chromosomes can be hundreds of millions of base pairs long, packaged into chromatin.

Another common macromolecular property that does not characterize smaller molecules is their relative insolubility in water and similar solvents, instead forming colloids. Many require salts or particular ions to dissolve in water. Similarly, many proteins will denature if the solute concentration of their solution is too high or too low.

High concentrations of macromolecules in a solution can alter the rates and equilibrium constants of the reactions of other macromolecules, through an effect known as macromolecular crowding.[11] This comes from macromolecules excluding other molecules from a large part of the volume of the solution, thereby increasing the effective concentrations of these molecules.

16.3   Linear biopolymers

All living organisms are dependent on three essential biopolymers for their biological functions: DNA, RNA and Proteins.[12] Each of these molecules is required for life since each plays a distinct, indispensable role in the cell.[13] The simple summary is that DNA makes RNA, and then RNA makes proteins.

DNA, RNA and proteins all consist of a repeating structure of related building blocks (nucleotides in the case of DNA and RNA, amino acids in the case of proteins). In general, they are all unbranched polymers, and so can be represented in the form of a string. Indeed, they can be viewed as a string of beads, with each bead representing a single nucleotide or amino acid monomer linked together through covalent chemical bonds into a very long chain.

In most cases, the monomers within the chain have a strong propensity to interact with other amino acids or nucleotides. In DNA and RNA, this can take the form of Watson-Crick base pairs (G-C and A-T or A-U), although many more complicated interactions can and do occur.

16.3.1   Structural features

Because of the double-stranded nature of DNA, essentially all of the nucleotides take the form of Watson-Crick base pairs between nucleotides on the two complementary strands of the double-helix.

In contrast, both RNA and proteins are normally single-stranded. Therefore, they are not constrained by the regular geometry of the DNA double helix, and so fold into complex three-dimensional shapes dependent on their sequence. These different shapes are responsible for many of the common properties of RNA and proteins, including the formation of specific binding pockets, and the ability to catalyse biochemical reactions.

DNA is optimised for encoding information

DNA is an information storage macromolecule that encodes the complete set of instructions (the genome) that are required to assemble, maintain, and reproduce every living organism.[14]

DNA and RNA are both capable of encoding genetic information, because there are biochemical mechanisms which read the information coded within a DNA or RNA sequence and use it to generate a specified protein. On the other hand, the sequence information of a protein molecule is not used by cells to functionally encode genetic information.[1]:5

DNA has three primary attributes that allow it to be far better than RNA at encoding genetic information. First, it is normally double-stranded, so that there are a minimum of two copies of the information encoding each gene in every cell. Second, DNA has a much greater stability against breakdown than does RNA, an attribute primarily associated with the absence of the 2'-hydroxyl group within every nucleotide of DNA. Third, highly sophisticated DNA surveillance and repair systems are present which monitor damage to the DNA and repair the sequence when necessary. Analogous systems have not evolved for repairing damaged RNA molecules. Consequently, chromosomes can contain many billions of atoms, arranged in a specific chemical structure.

Proteins are optimised for catalysis

Proteins are functional macromolecules responsible for catalysing the biochemical reactions that sustain life.[1]:3 Proteins carry out all functions of an organism, for example photosynthesis, neural function, vision, and movement.[15]

The single-stranded nature of protein molecules, together with their composition of 20 or more different amino acid building blocks, allows them to fold in to a vast number of different three-dimensional shapes, while providing binding pockets through which they can specifically interact with all manner of molecules. In addition, the chemical diversity of the different amino acids, together with different chemical environments afforded by local 3D structure, enables many proteins to act as Enzymes, catalyzing a wide range of specific biochemical transformations within cells. In addition, proteins have evolved the ability to bind a wide range of cofactors and Coenzymes, smaller molecules that can endow the protein with specific activities beyond those associated with the polypeptide chain alone.

RNA is multifunctional

RNA is multifunctional, its primary function is to encode proteins, according to the instructions within a cell's DNA.[1]:5 They control and regulate many aspects of protein synthesis in eukaryotes.

RNA encodes genetic information that can be translated into the amino acid sequence of proteins, as evidenced by the messenger RNA molecules present within every cell, and the RNA genomes of a large number of viruses. The single-stranded nature of RNA, together with tendency for rapid breakdown and a lack of repair systems means that RNA is not so well suited for the long-term storage of genetic information as is DNA.

In addition, RNA is a single-stranded polymer that can, like proteins, fold into a very large number of three-dimensional structures. Some of these structures provide binding sites for other molecules and chemically-active centers that can catalyze specific chemical reactions on those bound molecules. The limited number of different building blocks of RNA (4 nucleotides vs >20 amino acids in proteins), together with their lack of chemical diversity, results in catalytic RNA (ribozymes) being generally less-effective catalysts than proteins for most biological reactions.

16.4   Branched biopolymers

Raspberry ellagitannin, a tannin composed of core of glucose units surrounded by gallic acid esters and ellagic acid units

Carbohydrate macromolecules (polysaccharides) are formed from polymers of monosaccharides.[1]:11 Because monosaccharides have multiple functional groups, polysaccharides can form linear polymers (e.g. cellulose) or complex branched structures (e.g. glycogen). Polysaccharides perform numerous roles in living organisms, acting as energy stores (e.g. Starch) and as structural components (e.g.chitin in arthropods and fungi). Many carbohydrates contain modified monosaccharide units that have had functional groups replaced or removed.

Polyphenols consist of a branched structure of multiple phenolic subunits. They can perform structural roles (e.g. lignin) as well as roles as secondary metabolites involved in signalling, pigmentation and defense.

16.5 Synthetic macromolecules

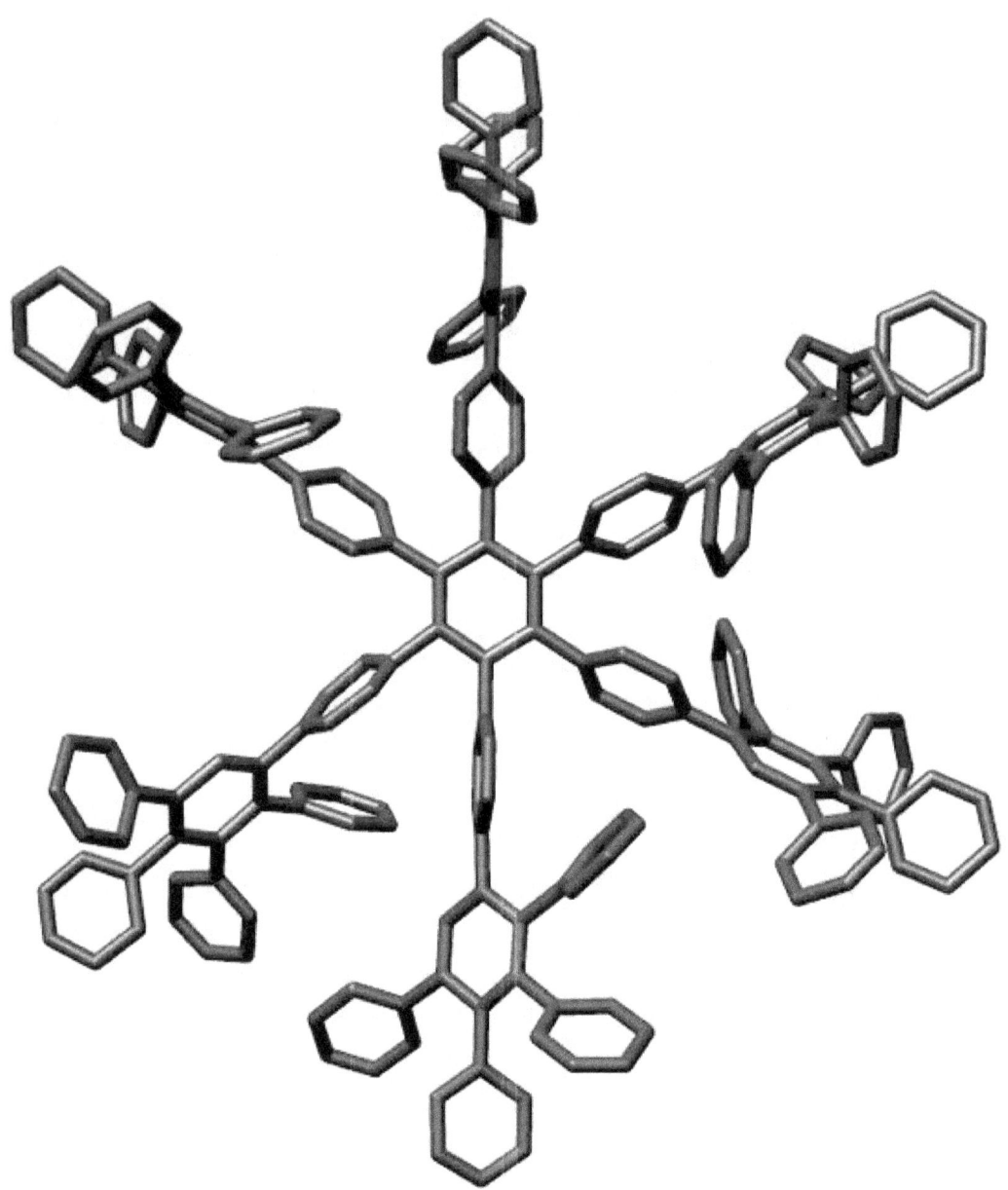

Structure of a polyphenylene dendrimer macromolecule reported by Müllen, et al.[16]

Some examples of macromolecules are synthetic polymers (plastics, synthetic fibers, and synthetic rubber), graphene, and carbon nanotubes.

16.6 References

[1] Stryer L, Berg JM, Tymoczko JL (2002). *Biochemistry* (5th ed.). San Francisco: W.H. Freeman. ISBN 0-7167-4955-6.

[2] Life cycle of a plastic product. Americanchemistry.com. Retrieved on 2011-07-01.

[3] Gullapalli, S.; Wong, M.S. (2011). "Nanotechnology: A Guide to Nano-Objects" (PDF). *Chemical Engineering Progress* **107** (5): 28–32.

[4] "Glossary of basic terms in polymer science (IUPAC Recommendations 1996)" (PDF). *Pure and Applied Chemistry* **68** (12): 2287–2311. 1996. doi:10.1351/pac199668122287.

[5] Staudinger, H.; Fritschi, J. (1922). "Über Isopren und Kautschuk. 5. Mitteilung. Über die Hydrierung des Kautschuks und über seine Konstitution". *Helvetica Chimica Acta* **5** (5): 785. doi:10.1002/hlca.19220050517.

[6]Jensen, William B. (2008). "The Origin of the Polymer Concept".*Journal of Chemical Education***85**(5): 624.Bibcode:2008JChE. doi:10.1021/ed085p624.

[7] van Holde, K.E. (1998) *Principles of Physical Biochemistry* Prentice Hall: New Jersey, ISBN 0-13-720459-0

[8] Jenkins, A. D.; Kratochvíl, P.; Stepto, R. F. T.; Suter, U. W. (1996). "Glossary of Basic Terms in Polymer Science" (PDF). *Pure and Applied Chemistry* **68** (12): 2287. doi:10.1351/pac199668122287.

[9] High Polymer Research Group

[10] Pauling, Linus (1964). *College Chemistry*. W.H. Feeman and Company.

[11] Minton AP (2006). "How can biochemical reactions within cells differ from those in test tubes?". *J. Cell. Sci.* **119** (Pt 14): 2863–9. doi:10.1242/jcs.03063. PMID 16825427.

[12] Berg, Jeremy Mark; Tymoczko, John L.; Stryer, Lubert (2010). *Biochemistry, 7th ed. (Biochemistry (Berg))*. W.H. Freeman & Company. ISBN 1-4292-2936-5. Fifth edition available online through the NCBI Bookshelf: link

[13] Walter, Peter; Alberts, Bruce; Johnson, Alexander S.; Lewis, Julian; Raff, Martin C.; Roberts, Keith (2008). *Molecular Biology of the Cell (5th edition, Extended version)*. New York: Garland Science. ISBN 0-8153-4111-3.. Fourth edition is available online through the NCBI Bookshelf: link

[14] Golnick, Larry; Wheelis, Mark. *The Cartoon Guide to Genetics*. Collins Reference. ISBN 978-0-06-273099-2.

[15] Takemura, Masaharu (2009). *The Manga Guide to Molecular Biology*. No Starch Press. ISBN 978-1-59327-202-9.

[16] Roland E. Bauer, Volker Enkelmann, Uwe M. Wiesler, Alexander J. Berresheim, Klaus Müllen (2002). "Single-Crystal Structures of Polyphenylene Dendrimers". *Chemistry – A European Journal* **8** (17): 3858. doi:10.1002/1521-3765(20020902)8:1-CHEM3858>3.0.CO;2-5.

- Tanford, Charles (1961). *Physical Chemistry of Macromolecules*. New York, NY: John Wiley & Sons.

16.7 External links

- Synopsis of Chapter 5, Campbell & Reece, 2002

- Lecture notes on the structure and function of macromolecules

- Several (free) introductory macromolecule related internet-based courses

- Giant Molecules! by Ulysses Magee, *ISSA Review* Winter 2002–2003, ISSN 1540-9864. Cached HTML version of a missing PDF file. Retrieved March 10, 2010. The article is based on the book, *Inventing Polymer Science: Staudinger, Carothers, and the Emergence of Macromolecular Chemistry* by Yasu Furukawa.

Chapter 17

Supermolecule

Carboxylic acid dimers.

The term **supermolecule** (or **supramolecule**) was introduced by Karl Lothar Wolf *et al.* (*Übermoleküle*) in 1937 to describe hydrogen-bonded acetic acid dimers.[1][2] The study of non-covalent association of complexes of molecules has since developed into the field of supramolecular chemistry. The term supermolecule is sometimes used to describe supramolecular assemblies, which are complexes of two or more molecules (often macromolecules) that are not covalently bonded.[3] [4] The term supermolecule is also used in biochemistry to describe complexes of biomolecules, such as peptides and oligonucleotides composed of multiple strands.[5]

17.1   See also

- Supramolecular chemistry

- Molecular self-assembly

- Supramolecular assembly

- Macromolecule

17.2   References

[1] Wolf, K. L., Frahm, H. Harms, H. (1937). "The State of Arrangement of Molecules in Liquids". *Z Phys. Chem., Abt. B*, **36**, 237–287.

[2] Historical Remarks on Supramolecular Chemistry – PDF (16 pg. paper)

[3] Supermolecule – thefreedictionary.com

[4] Lehn, Jean-Marie (1995). *Supramolecular Chemistry*. VCH. ISBN 3-527-29311-6.

[5] Lehninger, A. L. (1966). *Naturwiss.* 53, 57–63.

Chapter 18

List of compounds

For a large list of all (common) compounds, see the Dictionary of chemical formulas.

Compounds are organized into the following three lists:

- List of inorganic compounds, compounds without a C–H bond
- List of biomolecules.

18.1   See also

- Chemical compounds
- Chemical substance
- Inorganic compounds by element
- List of alloys
- List of alkanes
- List of elements by name
- List of minerals
- List of alchemical substances
- Polyatomic ion

18.2   External links

Relevant links for chemical compounds are:

- The CAS Substance Databases
- Common Chemistry
- Chemfinder
- ChemSpider

- ChemIDplus a non-commercial source

- PubChem(

- Quizlet - Long (long) list of chemical compounds

Chapter 19

Table of permselectivity for different substances

This is a table of permselectivity for different substances in the glomerulus of the kidney in renal filtration.

19.1 References

[1] Walter F. Boron, Emile L. Boulpaep. *Medical Physiology: A Cellular And Molecular Approach.* Elsevier/Saunders. ISBN 1-4160-2328-3. Page 761

Chapter 20

List of interstellar and circumstellar molecules

This is a list of molecules that have been detected in the interstellar medium and circumstellar envelopes, grouped by the number of component atoms. The chemical formula is listed for each detected compound, along with any ionized form that has also been observed.

20.1 Detection

The molecules listed below were detected by spectroscopy. Their spectral features are generated by transitions of component electrons between different energy levels, or by rotational or vibrational spectra. Detection usually occurs in radio, microwave, or infrared portions of the spectrum.[1]

Interstellar molecules are formed by chemical reactions within very sparse interstellar or circumstellar clouds of dust and gas. Usually this occurs when a molecule becomes ionized, often as the result of an interaction with a cosmic ray. This positively charged molecule then draws in a nearby reactant by electrostatic attraction of the neutral molecule's electrons. Molecules can also be generated by reactions between neutral atoms and molecules, although this process is generally slower.[2] The dust plays a critical role of shielding the molecules from the ionizing effect of ultraviolet radiation emitted by stars.[3]

20.1.1 History

The chemistry of life may have begun shortly after the Big Bang, 13.8 billion years ago, during a habitable epoch when the Universe was only 10–17 million years old.[4][5]

The first carbon-containing molecule detected in the interstellar medium was the methylidyne radical (CH) in 1937.[6] From the early 1970s it was becoming evident that interstellar dust consisted of a large component of more complex organic molecules (COMs),[7] probably polymers. Chandra Wickramasinghe proposed the existence of polymeric composition based on the molecule formaldehyde (H_2CO).[8] Fred Hoyle and Chandra Wickramasinghe later proposed the identification of bicyclic aromatic compounds from an analysis of the ultraviolet extinction absorption at 2175A.,[9] thus demonstrating the existence of polycyclic aromatic hydrocarbon molecules in space.

In 2004, scientists reported[10] detecting the spectral signatures of anthracene and pyrene in the ultraviolet light emitted by the Red Rectangle nebula (no other such complex molecules had ever been found before in outer space). This discovery was considered a confirmation of a hypothesis that as nebulae of the same type as the Red Rectangle approach the ends of their lives, convection currents cause carbon and hydrogen in the nebulae's core to get caught in stellar winds, and radiate outward.[11] As they cool, the atoms supposedly bond to each other in various ways and eventually form particles of a million or more atoms. The scientists inferred[10] that since they discovered polycyclic aromatic hydrocarbons (PAHs)

— which may have been vital in the formation of early life on Earth — in a nebula, by necessity they must originate in nebulae.[11]

In 2010, fullerenes (or "buckyballs") were detected in nebulae.[12] Fullerenes have been implicated in the origin of life; according to astronomer Letizia Stanghellini, "It's possible that buckyballs from outer space provided seeds for life on Earth."[13]

In October 2011, scientists found using spectroscopy that cosmic dust contains complex organic compounds ("amorphous organic solids with a mixed aromatic-aliphatic structure") that could be created naturally, and rapidly, by stars.[14][15][16] The compounds are so complex that their chemical structures resemble the makeup of coal and petroleum; such chemical complexity was previously thought to arise only from living organisms.[14] These observations suggest that organic compounds introduced on Earth by interstellar dust particles could serve as basic ingredients for life due to their surface-catalytic activities.[17][18] One of the scientists suggested that these compounds may have been related to the development of life on Earth and said that, "If this is the case, life on Earth may have had an easier time getting started as these organics can serve as basic ingredients for life."[14]

In August 2012, astronomers at Copenhagen University reported the detection of a specific sugar molecule, glycolaldehyde, in a distant star system. The molecule was found around the protostellarbinary *IRAS 16293-2422*, which is located 400 light years from Earth.[19][20] Glycolaldehyde is needed to form ribonucleic acid, or RNA, which is similar in function to DNA. This finding suggests that complex organic molecules may form in stellar systems prior to the formation of planets, eventually arriving on young planets early in their formation.[21]

In September 2012, NASA scientists reported that PAHs, subjected to interstellar medium (ISM) conditions, are transformed, through hydrogenation, oxygenation, and hydroxylation, to more complex organics — "a step along the path toward amino acids and nucleotides, the raw materials of proteins and DNA, respectively".[22][23] Further, as a result of these transformations, the PAHs lose their spectroscopic signature which could be one of the reasons "for the lack of PAH detection in interstellar ice grains, particularly the outer regions of cold, dense clouds or the upper molecular layers of protoplanetary disks."[22][23]

PAHs are found everywhere in deep space[24] and, in June 2013, PAHs were detected in the upper atmosphere of Titan, the largest moon of the planet Saturn.[25]

In 2013, Dwayne Heard at the University of Leeds suggested[26] that quantum mechanical tunneling could explain a reaction his group observed taking place, at a significantly higher than expected rate, between cold (around 63 Kelvin) hydroxyl and methanol molecules, apparently bypassing intramolecular energy barriers which would have to be overcome by thermal energy or ionization events for the same rate to exist at warmer temperatures. The proposed tunneling mechanism may help explain the common observation of fairly complex molecules (up to tens of atoms) in interstellar space.

A particularly large and rich region for detecting interstellar molecules is Sagittarius B2 (Sgr B2). This giant molecular cloud lies near the center of the Milky Way galaxy and is a frequent target for new searches. About half of the molecules listed below were first found near Sgr B2, and nearly every other molecule has since been detected in this feature.[27] A rich source of investigation for circumstellar molecules is the relatively nearby star CW Leonis (IRC +10216), where about 50 compounds have been identified.[28]

In March 2015, NASA scientists reported that, for the first time, complex DNA and RNA organic compounds of life, including uracil, cytosine and thymine, have been formed in the laboratory under outer space conditions, using starting chemicals, such as pyrimidine, found in meteorites. Pyrimidine, like polycyclic aromatic hydrocarbons (PAHs), the most carbon-rich chemical found in the Universe, may have been formed in red giants or in interstellar dust and gas clouds, according to the scientists.[29]

20.2 Molecules

The following tables list molecules that have been detected in the interstellar medium, grouped by the number of component atoms. If there is no entry in the Molecule column, only the ionized form has been detected. For molecules where no designation was given in the scientific literature, that field is left empty. Mass is given in Atomic mass units. The total number of unique species, including distinct ionization states, is listed in parentheses in each section header.

Most of the molecules detected so far are organic. Only one inorganic species has been observed in molecules which contain at least five atoms, SiH_4.[30] Larger molecules have so far all had at least one carbon atom, with no N-N or O-O bonds.[30]

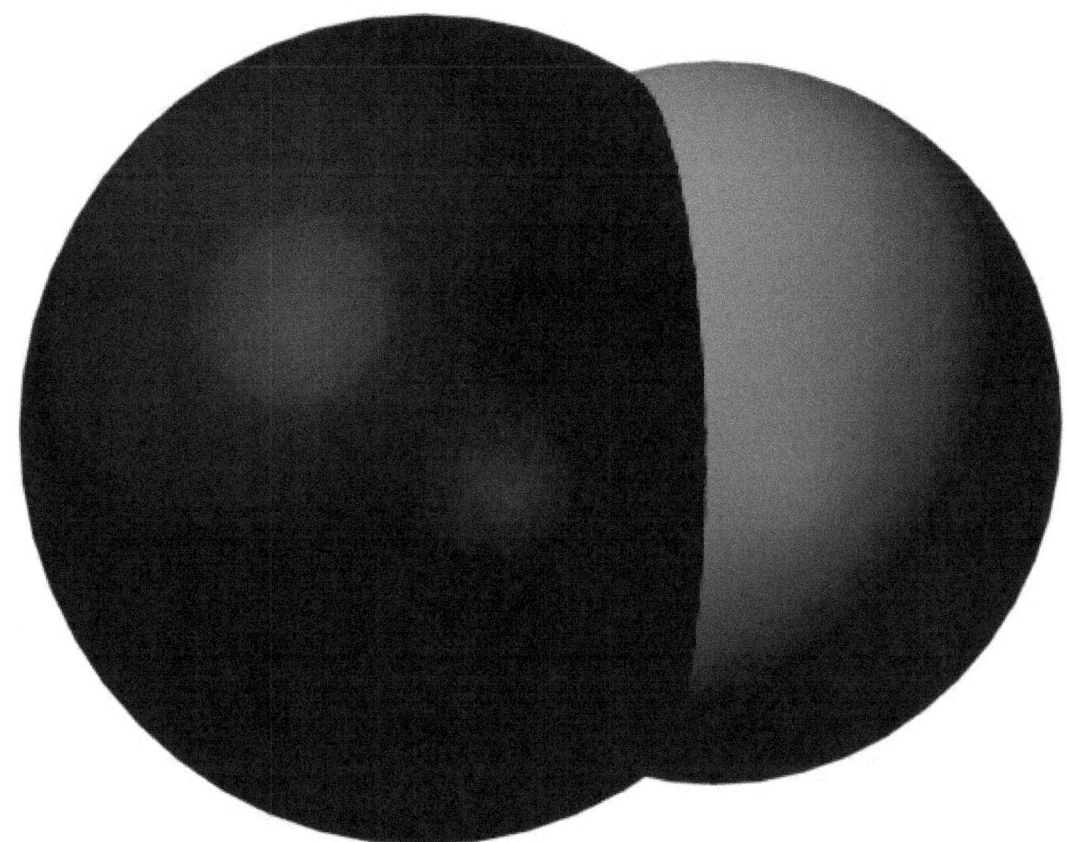

Carbon monoxide is frequently used to trace the distribution of mass in molecular clouds.[31]

The $H_3{}^+$ cation is one of the most abundant ions in the universe. It was first detected in 1993.[21][69]

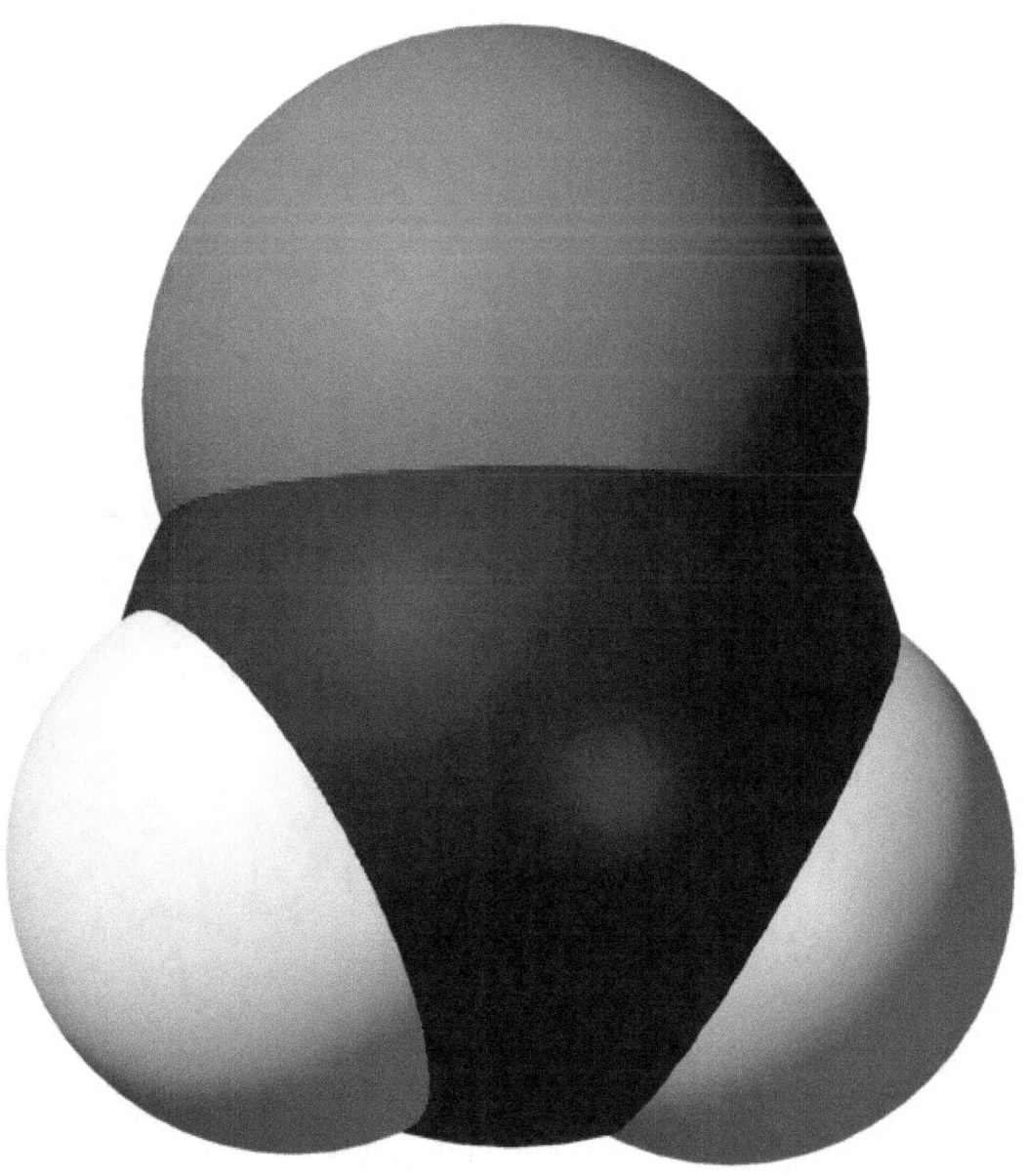

Formaldehyde is an organic molecule that is widely distributed in the interstellar medium.[98]

20.2.1   Diatomic (43)

20.2.2   Triatomic (43)

20.2.3   Four atoms (27)

20.2.4   Five atoms (18)

20.2.5   Six atoms (16)

20.2.6   Seven atoms (9)

20.2.7   Eight atoms (11)

20.2.8   Nine atoms (10)

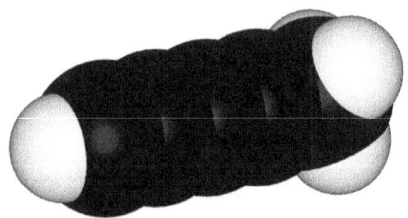

Methane, the primary component of natural gas, has also been detected on comets and in the atmosphere of several planets in the Solar System.[117]

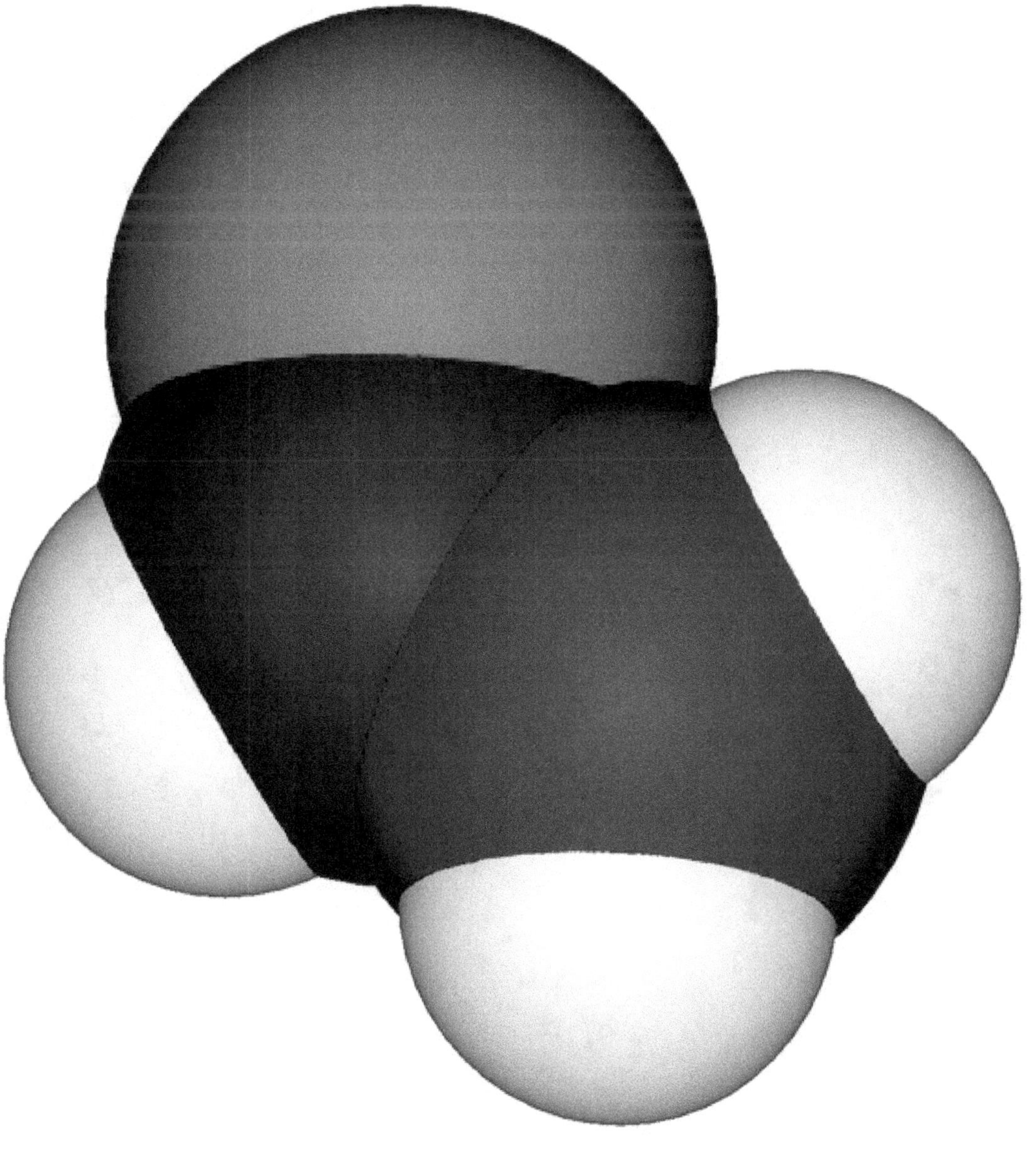

In the ISM, formamide (above) can combine with methylene to form acetamide.[132]

A number of polyyne-derived chemicals are among the heaviest molecules found in the interstellar medium.

Acetaldehyde (above) and its isomers vinyl alcohol and ethylene oxide have all been detected in interstellar space.[141]

20.2.9   Ten or more atoms (15)

20.3   Deuterated molecules (17)

These molecules all contain one or more deuterium atoms, a heavier isotope of hydrogen.

20.4   Unconfirmed (13)

Evidence for the existence of the following molecules has been reported in scientific literature, but the detections are either described as tentative by the authors, or have been challenged by other researchers. They await independent confirmation.

The radio signature of acetic acid, a compound found in vinegar, was confirmed in 1997.[145]

20.5 See also

- Abiogenesis

- Astrobiology

- Astrochemistry

- Atomic and molecular astrophysics

- Cosmic dust

- Cosmic ray

- Cosmochemistry

- Diffuse interstellar band

- Extraterrestrial liquid water

- Forbidden mechanism

- Intergalactic dust

- Interplanetary medium

- Interstellar medium

- Organic compound

- Outer space

- Panspermia

- Polycyclic aromatic hydrocarbon (PAH)

- Spectroscopy

20.6 References

[1] Shu, Frank H. (1982), *The Physical Universe: An Introduction to Astronomy*, University Science Books, ISBN 0-935702-05-9

[2] Dalgarno, A. (2006), "Interstellar Chemistry Special Feature: The galactic cosmic ray ionization rate", *Proceedings of the National Academy of Sciences* **103** (33): 12269–12273, Bibcode:2006PNAS..10312269D, doi:10.1073/pnas.0602117103, PMC 1567869, PMID 16894166

[3] Brown, Laurie M.; Pais, Abraham; Pippard, A. B. (1995), "The physics of the interstellar medium", *Twentieth Century Physics* (2nd ed.), CRC Press, p. 1765, ISBN 0-7503-0310-7

[4] Loeb, Abraham (October 2014). "The Habitable Epoch of the Early Universe". *International Journal of Astrobiology* **13** (04): 337–339. arXiv:1312.0613. Bibcode:2014IJAsB..13..337L. doi:10.1017/S1473550414000196. Retrieved 15 December 2014.

[5] Dreifus, Claudia (2 December 2014). "Much-Discussed Views That Go Way Back - Avi Loeb Ponders the Early Universe, Nature and Life". *New York Times*. Retrieved 3 December 2014.

[6] Woon, D. E. (May 2005), *Methylidyne radical*, The Astrochemist, retrieved 2007-02-13

[7] Ruaud, M.; Loison, J.C.; Hickson, K.M.; Gratier, P.; Hersant, F.; Wakelam, V. (22 December 2014). "Modeling Complex Organic Molecules in dense regions: Eley-Rideal and complex induced reaction" (PDF). *arXiv*. arXiv:1412.6256v1. Retrieved 26 January 2015.

[8] N.C. Wickramasinghe, Formaldehyde Polymers in Interstellar Space, Nature, 252, 462, 1974

[9] F. Hoyle and N.C. Wickramasinghe, Identification of the lambda 2200A interstellar absorption feature, Nature, 270, 323, 1977

[10] Battersby, S. (2004). "Space molecules point to organic origins". New Scientist. Retrieved 11 December 2009.

[11] Mulas, G.; Malloci, G.; Joblin, C.; Toublanc, D. (2006). "Estimated IR and phosphorescence emission fluxes for specific polycyclic aromatic hydrocarbons in the Red Rectangle". *Astronomy and Astrophysics* **446** (2): 537. arXiv:astro-ph/0509586. Bibcode:2006A&A...446..537M. doi:10.1051/0004-6361:20053738.

[12] García-Hernández, D. A.; Manchado, A.; García-Lario, P.; Stanghellini, L.; Villaver, E.; Shaw, R. A.; Szczerba, R.; Perea-Calderón, J. V. (2010-10-28). "Formation Of Fullerenes In H-Containing Planatary Nebulae". *The Astrophysical Journal Letters* **724** (1): L39–L43. arXiv:1009.4357. Bibcode:2010ApJ...724L..39G. doi:10.1088/2041-8205/724/1/L39.

[13] Atkinson, Nancy (2010-10-27). "Buckyballs Could Be Plentiful in the Universe". Universe Today. Retrieved 2010-10-28.

[14] Chow, Denise (26 October 2011). "Discovery: Cosmic Dust Contains Organic Matter from Stars". Space.com. Retrieved 2011-10-26.

[15] ScienceDaily Staff (26 October 2011). "Astronomers Discover Complex Organic Matter Exists Throughout the Universe". ScienceDaily. Retrieved 2011-10-27.

[16] Kwok, Sun; Zhang, Yong (26 October 2011). "Mixed aromatic–aliphatic organic nanoparticles as carriers of unidentified infrared emission features". *Nature* **479** (7371): 80–3. Bibcode:2011Natur.479...80K. doi:10.1038/nature10542. PMID 22031328.

[17] Gallori, Enzo (November 2010). "Astrochemistry and the origin of genetic material". *Rendiconti Lincei* **22** (2): 113–118. doi:10.1007/s12210-011-0118-4. Retrieved 2011-08-11.

[18] Martins, Zita (February 2011). "Organic Chemistry of Carbonaceous Meteorites". *Elements* **7** (1): 35–40. doi:10.2113/gse5. Retrieved 2011-08-11.

[19] Than, Ker (August 29, 2012). "Sugar Found In Space". *National Geographic*. Retrieved August 31, 2012.

[20] Staff (August 29, 2012). "Sweet! Astronomers spot sugar molecule near star". AP News. Retrieved August 31, 2012.

[21] Jørgensen, J. K.; Favre, C.; Bisschop, S.; Bourke, T.; Dishoeck, E.; Schmalzl, M. (2012). "Detection of the simplest sugar, gly-colaldehyde, in a solar-type protostar with ALMA" (PDF). *The Astrophysical Journal Letters*. eprint **757**: L4. arXiv:1208.5498. Bibcode:2012ApJ...757L...4J. doi:10.1088/2041-8205/757/1/L4.

[22] Staff (September 20, 2012). "NASA Cooks Up Icy Organics to Mimic Life's Origins". Space.com. Retrieved September 22, 2012.

[23] Gudipati, Murthy S.; Yang, Rui (September 1, 2012). "In-Situ Probing Of Radiation-Induced Processing Of Organics In Astrophysical Ice Analogs—Novel Laser Desorption Laser Ionization Time-Of-Flight Mass Spectroscopic Studies". *The Astrophysical Journal Letters* **756** (1). Bibcode:2012ApJ...756L..24G. doi:10.1088/2041-8205/756/1/L24. Retrieved September 22, 2012.

[24] Clavin, Whitney (10 February 2015). "Why Comets Are Like Deep Fried Ice Cream". *NASA*. Retrieved 10 February 2015.

[25] López-Puertas, Manuel (June 6, 2013). "PAH's in Titan's Upper Atmosphere". *CSIC*. Retrieved June 6, 2013.

[26] http://www.sciencenews.org/view/generic/id/351444/description/Interstellar_chemistry_makes_use_of_quantum_short 351468

[27] Cummins, S. E.; Linke, R. A.; Thaddeus, P. (1986), "A survey of the millimeter-wave spectrum of Sagittarius B2", *Astrophysical Journal Supplement Series* **60**: 819–878, Bibcode:1986ApJS...60..819C, doi:10.1086/191102

[28] Kaler, James B. (2002), *The hundred greatest stars*, Copernicus Series, Springer, ISBN 0-387-95436-8, retrieved 2011-05-09

[29] Marlaire, Ruth (3 March 2015). "NASA Ames Reproduces the Building Blocks of Life in Laboratory". *NASA*. Retrieved 5 March 2015.

[30] Klemperer, William (2011), "Astronomical Chemistry", *Annual Review of Physical Chemistry* **62**: 173–184, doi:10.1146/annur-physchem-032210-103332

[31] *The Structure of Molecular Cloud Cores*, Centre for Astrophysics and Planetary Science, University of Kent, retrieved 2007-02-16

[32] Ziurys, Lucy M. (2006), "The chemistry in circumstellar envelopes of evolved stars: Following the origin of the elements to the origin of life", *Proceedings of the National Academy of Sciences* **103** (33): 12274–12279, Bibcode:2006PNAS..10312274Z, doi:10.1073/pnas.0602277103, PMC 1567870, PMID 16894164

[33] Cernicharo, J.; Guelin, M. (1987), "Metals in IRC+10216 - Detection of NaCl, AlCl, and KCl, and tentative detection of AlF", *Astronomy and Astrophysics* **183** (1): L10–L12, Bibcode:1987A&A...183L..10C

[34] Ziurys, L. M.; Apponi, A. J.; Phillips, T. G. (1994), "Exotic fluoride molecules in IRC +10216: Confirmation of AlF and searches for MgF and CaF", *Astrophysical Journal* **433** (2): 729–732, Bibcode:1994ApJ...433..729Z, doi:10.1086/174682

[35] Tenenbaum, E. D.; Ziurys, L. M. (2009), "Millimeter Detection of AlO ($X^2\Sigma^+$): Metal Oxide Chemistry in the Envelope of VY Canis Majoris", *Astrophysical Journal* **694**: L59–L63, Bibcode:2009ApJ...694L..59T, doi:10.1088/0004-637X/694/1/L59

[36] Barlow, M. J.; Swinyard, B. M.; Owen, P. J.; Cernicharo, J.; Gomez, H. L.; Ivison, R. J.; Lim, T. L.; Matsuura, M.; Miller, S.; Olofsson, G.; Polehampton, E. T. (2013), "Detection of a Noble Gas Molecular Ion, ^{36}ArH+, in the Crab Nebula", *Science* **342** (6164): 1343–1345, doi:10.1126/science.124358213

[37] Quenqua, Douglas (13 December 2013). "Noble Molecules Found in Space". *New York Times*. Retrieved 13 December 2013.

[38] Lambert, D. L.; Sheffer, Y.; Federman, S. R. (1995), "Hubble Space Telescope observations of C_2 molecules in diffuse interstellar clouds", *Astrophysical Journal* **438**: 740–749, Bibcode:1995ApJ...438..740L, doi:10.1086/175119

[39] Galazutdinov, G. A.; Musaev, F. A.; Krelowski, J. (2001), "On the detection of the linear C_5 molecule in the interstellar medium", *Monthly Notices of the Royal Astronomical Society* **325** (4): 1332–1334, Bibcode:2001MNRAS.325.1332G, doi:10.1046/j.1365-8711.2001.04388.x

[40] Neufeld, D. A. et al. (2006), "Discovery of interstellar CF^+", *Astronomy and Astrophysics* **454** (2): L37–L40, arXiv:astro-ph/0603201, Bibcode:2006astro.ph..3201N, doi:10.1051/0004-6361:200600015

[41] Adams, Walter S. (1941), "Some Results with the COUDÉ Spectrograph of the Mount Wilson Observatory", *Astrophysical Journal* **93**: 11–23, Bibcode:1941ApJ....93...11A, doi:10.1086/144237

[42] Smith, D. (1988), "Formation and Destruction of Molecular Ions in Interstellar Clouds", *Philosophical Transactions of the Royal Society of London* **324** (1578): 257–273, Bibcode:1988RSPTA.324..257S, doi:10.1098/rsta.1988.0016

[43] Fuente, A. et al. (2005), "Photon-dominated Chemistry in the Nucleus of M82: Widespread HOC^+ Emission in the Inner 650 Parsec Disk", *Astrophysical Journal* **619** (2): L155–L158, arXiv:astro-ph/0412361, Bibcode:2005ApJ...619L.155F, doi:10.1086/427990

[44] Guelin, M.; Cernicharo, J.; Paubert, G.; Turner, B. E. (1990), "Free CP in IRC + 10216", *Astronomy and Astrophysics* **230**: L9–L11, Bibcode:1990A&A...230L...9G

[45] Dopita, Michael A.; Sutherland, Ralph S. (2003), *Astrophysics of the diffuse universe*, Springer-Verlag, ISBN 3-540-43362-7

[46] Agúndez, M. et al. (2010-07-30), "Astronomical identification of CN^-, the smallest observed molecular anion", *Astronomy & Astrophysics* **517**: L2, arXiv:1007.0662, Bibcode:2010A&A...517L...2A, doi:10.1051/0004-6361/201015186, retrieved 2010-09-03

[47] Khan, Amina. "Did two planets around nearby star collide? Toxic gas holds hints". *LA Times*. Retrieved March 9, 2014.

[48] Dent, W.R.F.; Wyatt, M.C.;Roberge, A.; Augereau,J.-C.; Casassus, S.;Corder, S.; Greaves, J.S.; de Gregorio-Monsalvo, I; Hales, A.; Jackson, A.P.; Hughes, A. Meredith; Lagrange, A.-M; Matthews, B.; Wilner, D. (March 6, 2014). "Molecular Gas Clumps from the Destruction of Icy Bodies in the β Pictoris Debris Disk". *Science*. arXiv:1404.1380. Bibcode:2014Sci...343.1 .doi:10.1126/science.1248726. Retrieved March 9, 2014.

[49] Latter, W. B.; Walker, C. K.; Maloney, P. R. (1993), "Detection of the Carbon Monoxide Ion (CO^+) in the Interstellar Medium and a Planetary Nebula", *Astrophysical Journal Letters* **419**: L97, Bibcode:1993ApJ...419L..97L, doi:10.1086/187146

[50] Furuya, R. S. et al. (2003), "Interferometric observations of FeO towards Sagittarius B2", *Astronomy and Astrophysics* **409** (2): L21–L24, Bibcode:2003A&A...409L..21F, doi:10.1051/0004-6361:20031304

[51] Adams, Walter S. (1970), "Rocket Observation of Interstellar Molecular Hydrogen", *Astrophysical Journal* **161**: L81–L85, Bibcode:1970ApJ...161L..81C, doi:10.1086/180575

[52] Blake, G. A.; Keene, J.; Phillips, T. G. (1985), "Chlorine in dense interstellar clouds - The abundance of HCl in OMC-1", *Astrophysical Journal, Part 1* **295**: 501–506, Bibcode:1985ApJ...295..501B, doi:10.1086/163394

[53] De Luca, M.; Gupta, H.; Neufeld, D.; Gerin, M.; Teyssier, D.; Drouin, B. J.; Pearson, J. C.; Lis, D. C. et al. (2012), "Herschel/HIFI Discovery of HCl+ in the Interstellar Medium", *The Astrophysical Journal Letters* **751** (2): L37, Bibcode: doi:10.1088/2041-8205/751/2/L37

[54] Neufeld, David A. et al. (1997), "Discovery of Interstellar Hydrogen Fluoride", *Astrophysical Journal Letters* **488** (2): L141–L144, arXiv:astro-ph/9708013, Bibcode:1997ApJ...488L.141N, doi:10.1086/310942

[55] Wyrowski, F. et al. (2009), "First interstellar detection of OH^+", *Astronomy & Astrophysics* **518**: A26, arXiv:1004.2627, Bibcode:2010A&A...518A..26W, doi:10.1051/0004-6361/201014364

[56] Meyer, D. M.; Roth, K. C. (1991), "Discovery of interstellar NH", *Astrophysical Journal, Part 2 - Letters* **376**: L49–L52, Bibcode:1991ApJ...376L..49M, doi:10.1086/186100

[57] Wagenblast, R. et al. (January 1993), "On the origin of NH in diffuse interstellar clouds", *Monthly Notices of the Royal Astronomical Society* **260** (2): 420–424, Bibcode:1993MNRAS.260..420W

[58] <Please add first missing authors to populate metadata.> (June 9, 2004), *Astronomers Detect Molecular Nitrogen Outside Solar System*, Space Daily, retrieved 2010-06-25

[59] Knauth, D. C et al. (2004), "The interstellar N_2 abundance towards HD 124314 from far-ultraviolet observations", *Nature* **429** (6992): 636–638, Bibcode:2004Natur.429..636K, doi:10.1038/nature02614, PMID 15190346, retrieved 2010-06-25

[60] McGonagle, D. et al. (1990), "Detection of nitric oxide in the dark cloud L134N", *Astrophysical Journal, Part 1* **359**: 121–124, Bibcode:1990ApJ...359..121M, doi:10.1086/169040

[61] Whiteoak, J. B.; Gardner, F. F. (1985), "Interstellar NaI absorption towards the stellar association ARA OB1", *Astronomical Society of Australia, Proceedings* (Sydney) **6** (2): 164–171, Bibcode:1985PASAu...6..164W

[62] Staff writers (March 27, 2007), *Elusive oxygen molecule finally discovered in interstellar space*, Physorg.com, retrieved 2007-04-02

[63] Ziurys, L. M. (1987), "Detection of interstellar PN - The first phosphorus-bearing species observed in molecular clouds", *Astrophysical Journal, Part 2 - Letters to the Editor* **321**: L81–L85, Bibcode:1987ApJ...321L..81Z, doi:10.1086/185010

[64] Tenenbaum, E. D.; Woolf, N. J.; Ziurys, L. M. (2007), "Identification of phosphorus monoxide (X ^{2} Pi $_r$) in VY Canis Majoris: Detection of the first PO bond in space", *Astrophysical Journal, Part 2 - Letters to the Editor* **666**: L29–L32, Bibcode:2007ApJ.. doi:10.1086/521361

[65] Yamamura, S. T.; Kawaguchi, K.; Ridgway, S. T. (2000), "Identification of SH v=1 Ro-vibrational Lines in R Andromedae", *The Astrophysical Journal* **528** (1): L33–L36, arXiv:astro-ph/9911080, Bibcode:2000ApJ...528L..33Y, doi:10.1086/312420, PMID 10587489

[66] Menten, K. M. et al. (2011), "Submillimeter Absorption from SH$^+$, a New Widespread Interstellar Radical, ^{13}CH$^+$ and HCl", *Astronomy & Astrophysics* **525**: A77, arXiv:1009.2825, Bibcode:2011A&A...525A..77M, doi:10.1051/0004-6361/201014363, retrieved 2010-12-03.

[67] Pascoli, G.; Comeau, M. (1995), "Silicon Carbide in Circumstellar Environment", *Astrophysics and Space Science* **226**: 149–163, Bibcode:1995Ap&SS.226..149P, doi:10.1007/BF00626907

[68] Kamiński, T. et al. (2013), "Pure rotational spectra of TiO and TiO_2 in VY Canis Majoris", *Astronomy and Astrophysics* **551**: A113, arXiv:1301.4344, Bibcode:2013A&A...551A.113K, doi:10.1051/0004-6361/201220290

[69]Geballe, T. R.; Oka, T. (1996), "Detection of H3+in Interstellar Space",*Nature***384**(6607): 334–335,Bibcode:1996Natur.384..3 34Gdoi:10.1038/384334a0, PMID 8934516

[70] Tenenbaum, E. D.; Ziurys, L. M. (2010), "Exotic Metal Molecules in Oxygen-rich Envelopes: Detection of AlOH (X1Σ+) in VY Canis Majoris", *Astrophysical Journal* **712**: L93–L97, Bibcode:2010ApJ...712L..93T, doi:10.1088/2041-8205/712/1/L93

[71] Anderson, J. K. et al. (2014), "Detection of CCN ($X^2\Pi_r$) in IRC+10216: Constraining Carbon-chain Chemistry", *Astrophysical Journal* **795**: L1, Bibcode:2014ApJ...795L...1A, doi:10.1088/2041-8205/795/1/L1

[72] Ohishi, Masatoshi, Masatoshi et al. (1991), "Detection of a new carbon-chain molecule, CCO", *Astrophysical Journal, Part 2 - Letters* **380**: L39–L42, Bibcode:1991ApJ...380L..39O, doi:10.1086/186168

[73] Irvine, William M. et al. (1988), "Newly detected molecules in dense interstellar clouds", *Astrophysical Letters and Communications* **26**: 167–180, Bibcode:1988ApL&C..26..167I, PMID 11538461

[74] Halfen, D. T.; Clouthier, D. J.; Ziurys, L. M. (2008), "Detection of the CCP Radical (X $^2\Pi_r$) in IRC +10216: A New Interstellar Phosphorus-containing Species", *Astrophysical Journal* **677** (2): L101–L104, Bibcode:2008ApJ...677L.101H,doi:10.1086/58

[75] Whittet, D. C. B.; Walker, H. J. (1991), "On the occurrence of carbon dioxide in interstellar grain mantles and ion-molecule chemistry", *Monthly Notices of the Royal Astronomical Society* **252**: 63–67, Bibcode:1991MNRAS.252...63W

[76] Zack, L. N.; Halfen, D. T.; Ziurys, L. M. (June 2011), "Detection of FeCN (X $^4\Delta_i$) in IRC+10216: A New Interstellar Molecule", *The Astrophysical Journal Letters* **733** (2): L36, Bibcode:2011ApJ...733L..36Z, doi:10.1088/2041-8205/733/2/L36

[77] Lis, D. C. et al. (2010-10-01), "Herschel/HIFI discovery of interstellar chloronium (H_2Cl^+)", *Astronomy & Astrophysics* **521**: L9, arXiv:1007.1461, Bibcode:2010A&A...521L...9L, doi:10.1051/0004-6361/201014959.

[78] *Europe's space telescope ISO finds water in distant places*, ESO, April 29, 1997, archived from the original on 2006-12-22, retrieved 2007-02-08

[79] Ossenkopf, V. et al. (2010), "Detection of interstellar oxidaniumyl: Abundant H2O$^+$ towards the star-forming regions DR21, Sgr B2, and NGC6334", *Astronomy & Astrophysics* **518**: L111, arXiv:1005.2521, Bibcode:2010A&A...518L.111O, doi:10.105 6361/201014577.

[80] Parise, B.; Bergman, P.; Du, F. (2012), "Detection of the hydroperoxyl radical HO$_2$ toward ρ Ophiuchi A. Additional constraints on the water chemical network", *Astronomy & Astrophysics Letters* **541**: L11–L14, arXiv:1205.0361, Bibcode:2012A&A...541L ,doi:10.1051/0004-6361/201219379

[81] Snyder, L. E.; Buhl, D. (1971), "Observations of Radio Emission from Interstellar Hydrogen Cyanide", *Astrophysical Journal* **163**: L47–L52, Bibcode:1971ApJ...163L..47S, doi:10.1086/180664

[82] Schilke, P.; Benford, D. J.; Hunter, T. R.; Lis, D. C., Phillips, T. G.; Phillips, T. G. (2001), "A Line Survey of Orion-KL from 607 to 725 GHz", *Astrophysical Journal Supplement Series* **132** (2): 281–364, Bibcode:2001ApJS..132..281S,doi:10.1086/3181

[83] Schenewerk, M. S.; Snyder, L. E.; Hjalmarson, A. (1986), "Interstellar HCO - Detection of the missing 3 millimeter quartet", *Astrophysical Journal, Part 2 - Letters to the Editor* **303**: L71–L74, Bibcode:1986ApJ...303L..71S, doi:10.1086/184655

[84] Kawaguchi, Kentarou et al. (1994), "Detection of a new molecular ion HC3NH(+) in TMC-1", *Astrophysical Journal* **420**: L95, Bibcode:1994ApJ...420L..95K, doi:10.1086/187171

[85] Agúndez, M.; Cernicharo, J.; Guélin, M. (2007), "Discovery of Phosphaethyne (HCP) in Space: Phosphorus Chemistry in Circumstellar Envelopes", *The Astrophysical Journal* **662** (2): L91, Bibcode:2007ApJ...662L..91A, doi:10.1086/519561, retrieved 2007-06-02

[86] Schilke, P.; Comito, C.; Thorwirth, S. (2003), "First Detection of Vibrationally Excited HNC in Space", *The Astrophysical Journal* **582** (2): L101–L104, Bibcode:2003ApJ...582L.101S, doi:10.1086/367628, retrieved 2008-09-14

[87] Hollis, J. M. et al. (1991), "Interstellar HNO: Confirming the Identification - Atoms, ions and molecules: New results in spectral line astrophysics", *Atoms* (San Francisco: ASP) **16**: 407–412, Bibcode:1991ASPC...16..407H

[88] van Dishoeck, Ewine F. et al. (1993), "Detection of the Interstellar NH 2 Radical", *Astrophysical Journal, Part 2 - Letters* **416**: L83–L86, Bibcode:1993ApJ...416L..83V, doi:10.1086/187076

[89] Womack, M.; Ziurys, L. M.; Wyckoff, S. (1992), "A survey of N$_2$H(+) in dense clouds - Implications for interstellar nitrogen and ion-molecule chemistry", *Astrophysical Journal, Part 1* **387**: 417–429, Bibcode:1992ApJ...387..417W, doi:10.1086/171094

[90] Ziurys, L. M. et al. (1994), "Detection of interstellar N$_2$O: A new molecule containing an N-O bond", *Astrophysical Journal, Part 2 - Letters* **436**: L181–L184, Bibcode:1994ApJ...436L.181Z, doi:10.1086/187662

[91] Hollis, J. M.; Rhodes, P. J. (November 1, 1982), "Detection of interstellar sodium hydroxide in self-absorption toward the galactic center", *Astrophysical Journal, Part 2 - Letters to the Editor* **262**: L1–L5, Bibcode:1982ApJ...262L...1H, doi:10.1086/183900

[92] Goldsmith, P. F.; Linke, R. A. (1981), "A study of interstellar carbonyl sulfide", *Astrophysical Journal, Part 1* **245**: 482–494, Bibcode:1981ApJ...245..482G, doi:10.1086/158824

[93] Phillips, T. G.; Knapp, G. R. (1980), "Interstellar Ozone",*American Astronomical Society Bulletin***12**: 440,Bibcode:1980BAAS...

[94] Johansson, L. E. B. et al. (1984), "Spectral scan of Orion A and IRC+10216 from 72 to 91 GHz", *Astronomy and Astrophysics* **130** (2): 227–256, Bibcode:1984A&A...130..227J

[95] Cernicharo, José et al. (2015), "Discovery of SiCSi in IRC+10216: a Missing Link Between Gas and Dust Carriers OF Si–C Bonds", *Astrophysical Journal Letters* **806**: L3, Bibcode:2015ApJ...806L.3C, doi:10.1088/2041-8025

[96] Guélin, M. et al. (2004), "Astronomical detection of the free radical SiCN", *Astronomy and Astrophysics* **363**: L9–L12, Bibcode:2000A&A...363L...9G

[97] Guélin, M. et al. (2004), "Detection of the SiNC radical in IRC+10216", *Astronomy and Astrophysics* **426** (2): L49–L52, Bibcode:2004A&A...426L..49G, doi:10.1051/0004-6361:200400074

[98] Snyder, Lewis E. et al. (1999), "Microwave Detection of Interstellar Formaldehyde", *Physical Review Letters* **61** (2): 77–115, Bibcode:1969PhRvL..22..679S, doi:10.1103/PhysRevLett.22.679

[99] Feuchtgruber, H. et al. (June 2000), "Detection of Interstellar CH$_3$", *The Astrophysical Journal* **535** (2): L111–L114, arXiv:astro-ph/0005273, Bibcode:2000ApJ...535L.111F, doi:10.1086/312711, PMID 10835311

[100] Irvine, W. M. et al. (1984), "Confirmation of the Existence of Two New Interstellar Molecules: C_3H and C_3O", *Bulletin of the American Astronomical Society* **16**: 877, Bibcode:1984BAAS...16..877I

[101] Pety, J. et al. (2012), "The IRAM-30 m line survey of the Horsehead PDR. II. First detection of the l-C_3MH^+ hydrocarbon cation", *Astronomy & Astrophysica* **548**: A68, arXiv:1210.8178, Bibcode:2012A&A...548A..68P, doi:10.1051/0004-6361/201220062

[102] Mangum, J. G.; Wootten, A. (1990), "Observations of the cyclic C_3H radical in the interstellar medium", *Astronomy and Astrophysics* **239**: 319–325, Bibcode:1990A&A...239..319M

[103] Wootten, Alwyn et al. (1991), "Detection of interstellar $H_3O(+)$ - A confirming line", *Astrophysical Journal, Part 2 - Letters* **380**: L79–L83, Bibcode:1991ApJ...380L..79W, doi:10.1086/186178

[104] Ridgway, S. T. et al. (1976), "Circumstellar acetylene in the infrared spectrum of IRC+10216", *Nature* **264**: 345, 346, Bibcode:1976Natur.264..345R, doi:10.1038/264345a0

[105] Ohishi, Masatoshi et al. (1994), "Detection of a new interstellar molecule, H_2CN", *Astrophysical Journal, Part 2 - Letters* **427**: L51–L54, Bibcode:1994ApJ...427L..51O, doi:10.1086/187362

[106] Minh, Y. C.; Irvine, W. M.; Brewer, M. K. (1991), "H2CS abundances and ortho-to-para ratios in interstellar clouds", *Astronomy and Astrophysics* **244**: 181–189, Bibcode:1991A&A...244..181M, PMID 11538284

[107] Guelin, M.; Cernicharo, J. (1991), "Astronomical detection of the HCCN radical - Toward a new family of carbon-chain molecules?", *Astronomy and Astrophysics* **244**: L21–L24, Bibcode:1991A&A...244L..21G

[108] Agúndez, M. et al. (2015), "Discovery of interstellar ketenyl (HCCO), a surprisingly abundant radical", *Astronomy and Astrophysics* **577**: A5, doi:10.1051/0004-6361

[109] Minh, Y. C.; Irvine, W. M.; Ziurys, L. M. (1988), "Observations of interstellar HOCO(+) - Abundance enhancements toward the Galactic center", *Astrophysical Journal, Part 1* **334**: 175–181, Bibcode:1988ApJ...334..175M, doi:10.1086/166827

[110] Marcelino, Núria et al. (2009), "Discovery of fulminic acid, HCNO, in dark clouds", *Astrophysical Journal* **690**: L27–L30, arXiv:0811.2679, Bibcode:2009ApJ...690L..27M, doi:10.1088/0004-637X/690/1/L27

[111] Brünken, S. et al. (2010-07-22), "Interstellar HOCN in the Galactic center region", *Astronomy & Astrophysics* **516**: A109, arXiv:1005.2489, Bibcode:2010A&A...516A.109B, doi:10.1051/0004-6361/200912456

[112] Bergman; Parise; Liseau; Larsson; Olofsson; Menten; Güsten (2011), "Detection of interstellar hydrogen peroxide", *Astronomy & Astrophysics* **531**: L8, arXiv:1105.5799, Bibcode:2011A&A...531L...8B, doi:10.1051/0004-6361/201117170.

[113] Frerking, M. A.; Linke, R. A.; Thaddeus, P. (1979), "Interstellar isothiocyanic acid", *Astrophysical Journal, Part 2 - Letters to the Editor* **234**: L143–L145, Bibcode:1979ApJ...234L.143F, doi:10.1086/183126

[114] Nguyen-Q-Rieu; Graham, D.; Bujarrabal, V. (1984), "Ammonia and cyanotriacetylene in the envelopes of CRL 2688 and IRC + 10216", *Astronomy and Astrophysics* **138** (1): L5–L8, Bibcode:1984A&A...138L...5N

[115] Halfen, D. T. et al. (September 2009), "Detection of a New Interstellar Molecule: Thiocyanic Acid HSCN", *The Astrophysical Journal Letters* **702** (2): L124–L127, Bibcode:2009ApJ...702L.124H, doi:10.1088/0004-637X/702/2/L124

[116] Cabezas, C. et al. (2013), "Laboratory and Astronomical Discovery of Hydromagnesium Isocyanide", *Astrophysical Journal* **775**: 133, arXiv:1309.0371, Bibcode:2013ApJ...775..133C, doi:10.1088/0004-637X/775/2/133

[117] Butterworth, Anna L. et al. (2004), "Combined element (H and C) stable isotope ratios of methane in carbonaceous chondrites", *Monthly Notices of the Royal Astronomical Society* **347** (3): 807–812, Bibcode:2004MNRAS.347..807B, doi:10.1111/j.1365-2966.2004.07251.x

[118] http://www.astro.uni-koeln.de/site/vorhersagen/molecules/ism/Ammonium.html

[119] http://iopscience.iop.org/2041-8205/771/1/L10/

[120] Cernicharo, J.; Marcelino, N.; Roueff, E.; Gerin, M.; Jiménez-Escobar, A.; Muñoz Caro, G. M. (2012), "Discovery of the Methoxy Radical, CH_3O, toward B1: Dust Grain and Gas-phase Chemistry in Cold Dark Clouds", *The Astrophysical Journal Letters* **759** (2): L43–L46, Bibcode:2012ApJ...759L..43C, doi:10.1088/2041-8205/759/2/L43

[121] Finley, Dave (August 7, 2006), *Researchers Use NRAO Telescope to Study Formation Of Chemical Precursors to Life*, National Radio Astronomy Observatory, retrieved 2006-08-10

[122] Fossé, David et al. (2001), "Molecular Carbon Chains and Rings in TMC-1", *Astrophysical Journal* **552** (1): 168–174, arXiv:astro-ph/0012405, Bibcode:2001ApJ...552..168F, doi:10.1086/320471, retrieved 2008-09-14

[123] Dickens, J. E. et al. (1997), "Hydrogenation of Interstellar Molecules: A Survey for Methylenimine (CH$_2$NH)", *Astrophysical Journal* **479** (1 Pt 1): 307–12, Bibcode:1997ApJ...479..307D, doi:10.1086/303884, PMID 11541227

[124] McGuire, B.A. et al. (2012), "Interstellar Carbodiimide (HNCNH): A New Astronomical Detection from the GBT PRIMOS Survey via Maser Emission Features", *The Astrophysical Journal Letters* **758** (2): L33–L38, arXiv:1209.1590, Bibcode:2012Ap, doi:10.1088/2041-8205/758/2/L33

[125] Ohishi, Masatoshi et al. (1996), "Detection of a New Interstellar Molecular Ion, H$_2$COH$^+$ (Protonated Formaldehyde)", *Astrophysical Journal* **471** (1): L61–4, Bibcode:1996ApJ...471L..61O, doi:10.1086/310325, PMID 11541244

[126] Cernicharo, J. et al. (2007), "Astronomical detection of C$_4$H<sup–, the second interstellar anion", *Astronomy and Astrophysics* **61** (2): L37–L40, Bibcode:2007A&A...467L..37C, doi:10.1051/0004-6361:20077415

[127] Walmsley, C. M.; Winnewisser, G.; Toelle, F. (1990), "Cyanoacetylene and cyanodiacetylene in interstellar clouds", *Astronomy and Astrophysics* **81** (1–2): 245–250, Bibcode:1980A&A....81..245W

[128] Kawaguchi, Kentarou et al. (1992), "Detection of isocyanoacetylene HCCNC in TMC-1", *Astrophysical Journal* **386** (2): L51–L53, Bibcode:1992ApJ...386L..51K, doi:10.1086/186290

[129] Turner, B. E. et al. (1975), "Microwave detection of interstellar cyanamide", *Astrophysical Journal* **201**: L149–L152, Bibcode: doi:10.1086/181963

[130] Remijan, Anthony J. et al. (2008), "Detection of interstellar cyanoformaldehyde (CNCHO)", *Astrophysical Journal* **675** (2): L85–L88, Bibcode:2008ApJ...675L..85R, doi:10.1086/533529

[131] Goldhaber, D. M.; Betz, A. L. (1984), "Silane in IRC +10216", *Astrophysical Journal, Part 2 - Letters to the Editor* **279**: –L55–L58, Bibcode:1984ApJ...279L..55G, doi:10.1086/184255

[132] Hollis, J. M. et al. (2006), "Detection of Acetamide (CH$_3$CONH$_2$): The Largest Interstellar Molecule with a Peptide Bond", *Astrophysical Journal* **643** (1): L25–L28, Bibcode:2006ApJ...643L..25H, doi:10.1086/505110

[133] Zaleski, D. P. et al. (2013), "Detection of E-Cyanomethanimine toward Sagittarius B2(N) in the Green Bank Telescope PRIMOS Survey", *Astrophysical Journal Letters* **765**: L109, arXiv:1302.0909, Bibcode:2013ApJ...765L..10Z, doi:10.1088/2041-8205/765/1/L10

[134] Betz, A. L. (1981), "Ethylene in IRC +10216", *Astrophysical Journal, Part 2 - Letters to the Editor* **244**: –L105, Bibcode:1981A, doi:10.1086/183490

[135] Remijan, Anthony J. et al. (2005), "Interstellar Isomers: The Importance of Bonding Energy Differences", *Astrophysical Journal* **632** (1): 333–339, arXiv:astro-ph/0506502, Bibcode:2005ApJ...632..333R, doi:10.1086/432908

[136] "Complex Organic Molecules Discovered in Infant Star System". *NRAO* (Astrobiology Web). 8 April 2015. Retrieved 2015-04-09.

[137] Cernicharo, José et al. (1997), "Infrared Space Observatory's Discovery of C$_4$H$_2$, C$_6$H$_2$, and Benzene in CRL 618", *Astrophysical Journal Letters* **546** (2): L123–L126, Bibcode:2001ApJ...546L.123C, doi:10.1086/318871

[138] Guelin, M.; Neininger, N.; Cernicharo, J. (1998), "Astronomical detection of the cyanobutadiynyl radical C_5N", *Astronomy and Astrophysics* **335**: L1–L4, arXiv:astro-ph/9805105, Bibcode:1998A&A...335L...1G

[139] Irvine, W. M. et al. (1988), "A new interstellar polyatomic molecule - Detection of propynal in the cold cloud TMC-1", *Astrophysical Journal, Part 2 - Letters* **335**: L89–L93, Bibcode:1988ApJ...335L..89I, doi:10.1086/185346

[140] Agúndez, M. et al. (2014), "New molecules in IRC +10216: confirmation of C$_5$S and tentative identification of MgCCH, NCCP, and SiH$_3$CN", *Astronomy and Astrophysics* **570**: A45, doi:10.1051/0004-6361

[141] *Scientists Toast the Discovery of Vinyl Alcohol in Interstellar Space*, National Radio Astronomy Observatory, October 1, 2001, retrieved 2006-12-20

[142] Dickens, J. E. et al. (1997), "Detection of Interstellar Ethylene Oxide (c-C2H4O)", *The Astrophysical Journal* **489** (2): 753–757, Bibcode:1997ApJ...489..753D, doi:10.1086/304821, PMID 11541726

[143] Kaifu, N.; Takagi, K.; Kojima, T. (1975), "Excitation of interstellar methylamine", *Astrophysical Journal* **198**: L85–L88, Bibcode:1975ApJ...198L..85K, doi:10.1086/181818

[144] McCarthy, M. C. et al. (2006), "Laboratory and Astronomical Identification of the Negative Molecular Ion C_6H^-", *Astrophysical Journal* **652** (2): L141–L144, Bibcode:2006ApJ...652L.141M, doi:10.1086/510238

[145] Mehringer, David M. et al. (1997), "Detection and Confirmation of Interstellar Acetic Acid", *Astrophysical Journal Letters* **480**: L71, Bibcode:1997ApJ...480L..71M, doi:10.1086/310612

[146] Lovas, F. J. et al. (2006), "Hyperfine Structure Identification of Interstellar Cyanoallene toward TMC-1", *Astrophysical Journal Letters* **637** (1): L37–L40, Bibcode:2006ApJ...637L..37L, doi:10.1086/500431

[147] Sincell, Mark (June 27, 2006), *The Sweet Signal of Sugar in Space*, American Association for the Advancement of Science, retrieved 2006-12-20

[148] Loomis, R. A. et al. (2013), "The Detection of Interstellar Ethanimine CH_3CHNH) from Observations Taken during the GBT PRIMOS Survey", *Astrophysical Journal Letters* **765**: L9, arXiv:1302.1121, Bibcode:2013ApJ...765L...9L, doi:10.1088/2041-8205/765/1/L9

[149] Guelin, M. et al. (1997), "Detection of a new linear carbon chain radical: C_7H", *Astronomy and Astrophysics* **317**: L37–L40, Bibcode:1997A&A...317L...1G

[150] Belloche, A. et al. (2008), "Detection of amino acetonitrile in Sgr B2(N)", *Astronomy & Astrophysics* **482**: 179–196, arXiv:0801., Bibcode:2008A&A...482..179B, doi:10.1051/0004-6361:20079203

[151] Remijan, Anthony J. et al. (2014), "OBSERVATIONAL RESULTS OF A MULTI-TELESCOPE CAMPAIGN IN SEARCH OF INTERSTELLAR UREA [(NH2)$_2$CO]", *Astrophysical Journal* **783** (2): 77, arXiv:1401.4483, Bibcode:2014ApJ...783...77R, doi:10.1088/0004-637X/783/2/77

[152] Remijan, Anthony J. et al. (2006), "Methyltriacetylene (CH_3C_6H) toward TMC-1: The Largest Detected Symmetric Top", *Astrophysical Journal* **643** (1): L37–L40, Bibcode:2006ApJ...643L..37R, doi:10.1086/504918

[153] Snyder, L. E. et al. (1974), "Radio Detection of Interstellar Dimethyl Ether", *Astrophysical Journal* **191**: L79–L82, Bibcode: S, doi:10.1086/181554

[154] Zuckerman, B. et al. (1975), "Detection of interstellar trans-ethyl alcohol", *Astrophysical Journal* **196** (2): L99–L102, Bibcode: doi:10.1086/181753

[155] Cernicharo, J.; Guelin, M. (1996), "Discovery of the C_8H radical", *Astronomy and Astrophysics* **309**: L26–L30, Bibcode:1996A

[156] Brünken, S. et al. (2007), "Detection of the Carbon Chain Negative Ion C_8H^- in TMC-1", *Astrophysical Journal* **664** (1): L43–L46, Bibcode:2007ApJ...664L..43B, doi:10.1086/520703

[157] Bell, M. B. et al. (1997), "Detection of $HC_{11}N$ in the Cold Dust Cloud TMC-1", *Astrophysical Journal Letters* **483** (1): L61–L64, arXiv:astro-ph/9704233, Bibcode:1997ApJ...483L..61B, doi:10.1086/310732

[158] Kroto, H. W. et al. (1978), "The detection of cyanohexatriyne, H$(C\equiv C)_3CN$, in Heiles's cloud 2", *The Astrophysical Journal* **219**: L133–L137, Bibcode:1978ApJ...219L.133K, doi:10.1086/182623

[159] Marcelino, N. et al. (2007), "Discovery of Interstellar Propylene (CH_2CHCH_3): Missing Links in Interstellar Gas-Phase Chemistry", *Astrophysical Journal* **665** (2): L127–L130, arXiv:0707.1308, Bibcode:2007ApJ...665L.127M, doi:10.1086/521398

[160] Kolesniková, L. et al. (2014), "Spectroscopic Characterization and Detection of Ethyl Mercaptan in Orion", *Astrophysical Journal Letters* **784** (1): L7, arXiv:1401.7810, Bibcode:2014ApJ...784L...7K, doi:10.1088/2041-8205/784/1/L7

[161] Snyder, Lewis E. et al. (2002), "Confirmation of Interstellar Acetone", *The Astrophysical Journal* **578** (1): 245–255, Bibcode: doi:10.1086/342273

[162] Hollis, J. M. et al. (2002), "Interstellar Antifreeze: Ethylene Glycol", *Astrophysical Journal* **571** (1): L59–L62, Bibcode:20, doi:10.1086/341148, retrieved 2010-07-18

[163] Hollis, J. M. (2005), "Complex Molecules and the GBT: Is Isomerism the Key?" (PDF), *Complex Molecules and the GBT: Is Isomerism the Key?*, Proceedings of the IAU Symposium 231, Astrochemistry throughout the Universe, Asilomar, CA, pp. 119–127

[164] Eyre, Michael (26 September 2014). "Complex organic molecule found in interstellar space". *BBC News*. Retrieved 2014-09-26.

[165] Belloche, Arnaud; Garrod, Robin T.; Müller, Holger S. P.; Menten, Karl M. (26 September 2014). "Detection of a branched alkyl molecule in the interstellar medium: iso-propyl cyanide". *Science* **345** (6204): 1584–1587. arXiv:1410.2607. Bibcode: .doi:10.1126/science.1256678. Retrieved 2014-09-26.

[166] Belloche, A. et al. (May 2009), "Increased complexity in interstellar chemistry: Detection and chemical modeling of ethyl formate and n-propyl cyanide in Sgr B2(N)", *Astronomy and Astrophysics* **499** (1): 215–232, arXiv:0902.4694, Bibcode:2009A&A. ,doi:10.1051/0004-6361/200811550

[167] Tercero, B. et al. (2013), "Discovery of Methyl Acetate and Gauche Ethyl Formate in Orion", *Astrophysical Journal Letters* **770**: L13, arXiv:1305.1135, Bibcode:2013ApJ...770L..13T, doi:10.1088/2041-8205/770/1/L13

[168] Cami, Jan et al. (July 22, 2010), "Detection of C_{60} and C_{70} in a Young Planetary Nebula", *Science* **329** (5996): 1180–2, Bibcode:2010Sci...329.1180C, doi:10.1126/science.1192035, PMID 20651118

[169] Foing, B. H.; Ehrenfreund, P. (1994), "Detection of two interstellar absorption bands coincident with spectral features of C60+", *Nature* **369** (6478): 296, Bibcode:1994Natur.369..296F, doi:10.1038/369296a0.

[170] Berné, Olivier; Mulas, Giacomo; Joblin, Christine (2013), "Interstellar C_{60+}",*Astronomy & Astrophysics***550**: L4,arXiv:1211.72, Bibcode:2013A&A...550L...4B, doi:10.1051/0004-6361/201220730

[171] Lacour, S. et al. (2005), "Deuterated molecular hydrogen in the Galactic ISM. New observations along seven translucent sightlines", *Astronomy and Astrophysics* **430** (3): 967–977, arXiv:astro-ph/0410033,Bibcode:2005A&A...430..967L,doi:10.1051/00-6361:20041589

[172] Ceccarelli, Cecilia (2002), "Millimeter and infrared observations of deuterated molecules", *Planetary and Space Science* **50** (12–13): 1267–1273, Bibcode:2002P&SS...50.1267C, doi:10.1016/S0032-0633(02)00093-4

[173] Green, Sheldon (1989), "Collisional excitation of interstellar molecules - Deuterated water, HDO", *Astrophysical Journal Supplement Series* **70**: 813–831, Bibcode:1989ApJS...70..813G, doi:10.1086/191358

[174] Butner, H. M. et al. (2007), "Discovery of interstellar heavy water", *Astrophysical Journal* **659** (2): L137–L140, Bibcode:20, doi:10.1086/517883

[175] Turner, B. E.; Zuckerman, B. (1978), "Observations of strongly deuterated molecules - Implications for interstellar chemistry", *Astrophysical Journal, Part 2 - Letters to the Editor* **225**: L75–L79, Bibcode:1978ApJ...225L..75T, doi:10.1086/182797

[176] Lis, D. C. et al. (2002), "Detection of Triply Deuterated Ammonia in the Barnard 1 Cloud", *Astrophysical Journal* **571** (1): L55–L58, Bibcode:2002ApJ...571L..55L, doi:10.1086/341132.

[177] Hatchell, J. (2003), "High NH_2D/NH_3 ratios in protostellar cores", *Astronomy and Astrophysics* **403** (2): L25–L28, arXiv:astro-ph/0302564, Bibcode:2003A&A...403L..25H, doi:10.1051/0004-6361:20030297.

[178] Turner, B. E. (1990), "Detection of doubly deuterated interstellar formaldehyde (D2CO) - an indicator of active grain surface chemistry", *Astrophysical Journal, Part 2 - Letters* **362**: L29–L33, Bibcode:1990ApJ...362L..29T, doi:10.1086/185840.

[179] Cernicharo, J. et al. (2013), "Detection of the Ammonium ion in space", *Astrophysical Journal Letters* **771**: L10, arXiv:1306.3364, Bibcode:2013ApJ...771L..10C, doi:10.1088/2041-8205/771/1/L10

[180] Doménech, J. L. et al. (2013), "Improved Determinination of the 1_0-0_0 Rotational Frequency of NH_3D^+ from the High-Resolution Spectrum of the ν_4 Infrared Band", *Astrophysical Journal Letters* **771**: L11, arXiv:1306.3792, Bibcode:2013ApJ..., doi:10.1088/2041-8205/771/1/L10

[181] Gerin, M. et al. (1992), "Interstellar detection of deuterated methyl acetylene", *Astronomy and Astrophysics* **253** (2): L29–L32, Bibcode:1992A&A...253L..29G.

[182] Markwick, A. J.; Charnley, S. B.; Butner, H. M.; Millar, T. J. (2005), "Interstellar CH3CCD", *The Astrophysical Journal* **627** (2): L117–L120, Bibcode:2005ApJ...627L.117M, doi:10.1086/432415.

[183] Agúndez, M. et al. (2008-06-04), "Tentative detection of phosphine in IRC +10216", *Astronomy & Astrophysics* **485** (3): L33, arXiv:0805.4297, Bibcode:2008A&A...485L..33A, doi:10.1051/0004-6361:200810193

[184] Gupta, H. et al. (2013), "Laboratory Measurements and Tentative Astronomical Identification of H_2NCO^+", *Astrophysical Journal Letters* **778**: L1, Bibcode:2013ApJ...778L...1G, doi:10.1088/2041-8205/778/1/L1

[185] Kuan, Y. J. et al. (2003), "Interstellar Glycine", *Astrophysical Journal* **593** (2): 848–867, Bibcode:2003ApJ...593..848K, doi:10.1086/375637.

[186] Widicus Weaver, S. L.; Blake, G. A. (2005), "1,3-Dihydroxyacetone in Sagittarius B2(N-LMH): The First Interstellar Ketose", *Astrophysical Journal Letters* **624** (1): L33–L36, Bibcode:2005ApJ...624L..33W, doi:10.1086/430407

[187] Fuchs, G. W. et al. (2005), "Trans-Ethyl Methyl Ether in Space: A new Look at a Complex Molecule in Selected Hot Core Regions", *Astronomy & Astrophysics* **444** (2): 521–530, arXiv:astro-ph/0508395,Bibcode:2005A&A...444..521F,doi:10.1051/0 6361:20053599, retrieved 2010-07-18

[188] Iglesias-Groth, S. et al. (2008-09-20), "Evidence for the Naphthalene Cation in a Region of the Interstellar Medium with Anomalous Microwave Emission", *The Astrophysical Journal Letters* **685**: L55–L58, arXiv:0809.0778, Bibcode:2008ApJ...6, doi:10.1086/592349 - This spectral assignment has not been independently confirmed, and is described by the authors as "tentative" (page L58).

[189] García-Hernández, D. A. et al. (2011), "The Formation of Fullerenes: Clues from New C_{60}, C_{70}, and (Possible) Planar C_{24} Detections in Magellanic Cloud Planetary Nebulae", *Astrophysical Journal Letters* **737** (2): L30, arXiv:1107.2595, Bibcode: doi:10.1088/2041-8205/737/2/L30, retrieved 2011-08-12.

[190] Iglesias-Groth, S. et al. (May 2010), "A search for interstellar anthracene toward the Perseus anomalous microwave emission region", *Monthly Notices of the Royal Astronomical Society* **407** (4): 2157, arXiv:1005.4388, Bibcode:2010MNRAS.407.2157I, doi:10.1111/j.1365-2966.2010.17075.x

20.7 External links

- Woon, David E. (October 1, 2010). "Interstellar and Circumstellar Molecules". Retrieved 2010-10-04.

- "Molecules in Space". Universität zu Köln. August 2010. Retrieved 2010-10-04.

- Dworkin, Jason P. (February 1, 2007). "Interstellar Molecules". NASA's Cosmic Ice Lab. Retrieved 2010-12-23.

- Wootten, Al (November 2005). "The 129 reported interstellar and circumstellar molecules". National Radio Astronomy Observatory. Retrieved 2007-02-13.

- Lovas, F. J.; Dragoset, R. A. (February 2004). "NIST Recommended Rest Frequencies for Observed Interstellar Molecular Microwave Transitions, 2002 Revision". National Institute of Standards and Technology. Retrieved 2007-02-13.

Chapter 21

List of software for molecular mechanics modeling

This is a list of computer programs that are predominantly used for **molecular mechanics** calculations.

Min - Optimization, **MD** - Molecular Dynamics, **MC** - Monte Carlo, **REM** - Replica exchange method, **QM** - Quantum mechanics, **Imp** - Implicit water, **GPU** - GPU accelerated.

Y - Yes.
I - Has interface.

21.1 See also

- Molecular dynamics

- Molecular design software

- Molecule editor

- Molecular modeling on GPUs

- List of molecular graphics systems

- List of nucleic acid simulation software

- List of protein structure prediction software

- List of quantum chemistry and solid state physics software

- List of software for nanostructures modeling

- List of software for Monte Carlo molecular modeling

- Force field implementation

21.2 Notes and references

[1] M. J. Harvey, G. Giupponi and G. De Fabritiis (2009). "ACEMD: Accelerating Biomolecular Dynamics in the Microsecond Time Scale". *Journal of Chemical Theory and Computation* **5** (6): 1632–1639. doi:10.1021/ct9000685.

[2] Johnston, MA, Fernández-Galván, I, Villà-Freixa, J (2005). "Framework-based design of a new all-purpose molecular simulation application: the Adun simulator". *J. Comp. Chem.* **26** (15): 1647–1659. doi:10.1002/jcc.20312. PMID 16175583.

[3] Cornell WD, Cieplak P, Bayly CI, Gould IR, Merz KM Jr, Ferguson DM, Spellmeyer DC, Fox T, Caldwell JW, Kollman PA (1995). "A second generation force field for the simulation of proteins, nucleic acids, and organic molecules". *J. Am. Chem. Soc.* **117** (19): 5179–5197. doi:10.1021/ja00124a002.

[4] Implicit Solvent - Gromacs

[5] Macke T, Case DA (1998). "Modeling unusual nucleic acid structures". *Molecular Modeling of Nucleic Acids*: 379–393.

[6] A. Górecki, M. Szypowski, M. Długosz and J. Trylska (2009). "RedMD - Reduced molecular dynamics package". *J. Comp. Chem.* **30** (14): 2364–2373. doi:10.1002/jcc.21223. PMID 19247989.

21.3 External links

- SINCRIS

- Linux4Chemistry

- Collaborative Computational Project

- World Index of Molecular Visualization Resources

- Short list of Molecular Modeling resources

- OpenScience

- Biological Magnetic Resonance Data Bank

- Materials modelling and computer simulation codes

- A few tips on molecular dynamics

Chapter 22

Molecular design software

Molecular design software is software for molecular modeling, that provides special support for developing molecular models *de novo*.

In contrast to the usual molecular modeling programs such as the molecular dynamics and quantum chemistry programs, such software directly supports the aspects related to the construction of molecular models:

- Molecular graphics

- interactive molecular drawing and conformational editing

- building of polymeric molecules, crystals and solvated systems

- partial charges development

- geometry optimization

- support for the different aspects of Force Field development

- *etc.*

22.1   Comparative table of packages covering the major aspects of molecular design

3D - Molecular Graphics, **Mouse** - drawing molecule by mouse, **Poly** - polymer building, **DNA** - Nucleic acid building, **Pept** - Peptide building, **Cryst** - crystal building, **Solv** - solvent addition, **Q** - partial charges, **Dock** - docking, **Min** - optimization, **MM** - Molecular mechanics, **QM** - Quantum mechanics. **FF** - Support for Force Field development. **QSAR** - 2D, 3D and Group QSAR.

22.2   Notes and references

22.3   See also

- Molecule editor

- Molecular modelling

- Molecular modelling on GPU

- Protein design

- Drug design

- Force field

- Force field implementation

- Nucleic acid simulation software

- List of molecular graphics systems

- Software for molecular mechanics modeling

- List of software for Monte Carlo molecular modeling

- List of software for nanostructures modeling

- Quantum chemistry computer programs

- Quantitative structure-activity relationship

22.4   External links

- molecular design *IUPAC* term definition.

- Journal of Computer-Aided Molecular Design

- Molecular Modeling resources

- Materials modelling and computer simulation codes

- Click2Drug.org Directory of in silico (computer-aided) drug design tools.

Chapter 23

Molecular modelling

Molecular modelling encompasses all theoretical methods and computational techniques used to model or mimic the behaviour of molecules. The techniques are used in the fields of computational chemistry, drug design, computational biology and materials science for studying molecular systems ranging from small chemical systems to large biological molecules and material assemblies. The simplest calculations can be performed by hand, but inevitably computers are required to perform molecular modelling of any reasonably sized system. The common feature of molecular modelling techniques is the atomistic level description of the molecular systems. This may include treating atoms as the smallest individual unit (the Molecular mechanics approach), or explicitly modeling electrons of each atom (the quantum chemistry approach).

23.1   Molecular mechanics

Molecular mechanics is one aspect of molecular modelling, as it refers to the use of classical mechanics/Newtonian mechanics to describe the physical basis behind the models. Molecular models typically describe atoms (nucleus and electrons collectively) as point charges with an associated mass. The interactions between neighbouring atoms are described by spring-like interactions (representing chemical bonds) and van der Waals forces. The Lennard-Jones potential is commonly used to describe van der Waals forces. The electrostatic interactions are computed based on Coulomb's law. Atoms are assigned coordinates in Cartesian space or in internal coordinates, and can also be assigned velocities in dynamical simulations. The atomic velocities are related to the temperature of the system, a macroscopic quantity. The collective mathematical expression is known as a potential function and is related to the system internal energy (U), a thermodynamic quantity equal to the sum of potential and kinetic energies. Methods which minimize the potential energy are known as energy minimization techniques (e.g., steepest descent and conjugate gradient), while methods that model the behaviour of the system with propagation of time are known as molecular dynamics.

$$E = E_{\text{bonds}} + E_{\text{angle}} + E_{\text{dihedral}} + E_{\text{non-bonded}}$$

$$E_{\text{non-bonded}} = E_{\text{electrostatic}} + E_{\text{Waals der van}}$$

This function, referred to as a potential function, computes the molecular potential energy as a sum of energy terms that describe the deviation of bond lengths, bond angles and torsion angles away from equilibrium values, plus terms for non-bonded pairs of atoms describing van der Waals and electrostatic interactions. The set of parameters consisting of equilibrium bond lengths, bond angles, partial charge values, force constants and van der Waals parameters are collectively known as a force field. Different implementations of molecular mechanics use different mathematical expressions and different parameters for the potential function. The common force fields in use today have been developed by using high level quantum calculations and/or fitting to experimental data. The technique known as energy minimization is used to find positions of zero gradient for all atoms, in other words, a local energy minimum. Lower energy states are more stable and are commonly investigated because of their role in chemical and biological processes. A molecular dynamics simulation, on the other hand, computes the behaviour of a system as a function of time. It involves solving Newton's laws of motion, principally the second law, $\mathbf{F} = m\mathbf{a}$. Integration of Newton's laws of motion, using different integration algorithms, leads to atomic trajectories in space and time. The force on an atom is defined as the negative gradient of the potential energy function. The energy minimization technique is useful for obtaining a static picture for comparing between states of similar systems, while molecular dynamics provides information about the dynamic processes with the intrinsic inclusion of temperature effects.

23.2   Variables

Molecules can be modeled either in vacuum or in the presence of a solvent such as water. Simulations of systems in vacuum are referred to as *gas-phase* simulations, while those that include the presence of solvent molecules are referred to as *explicit solvent* simulations. In another type of simulation, the effect of solvent is estimated using an empirical mathematical expression; these are known as *implicit solvation* simulations.

23.2.1   Coordinate Representations

Most force fields are distance dependent, making the most convenient expression for these Cartesian coordinates. Yet the comparatively rigid nature of bonds which occur between specific atoms, and in essence, defines what we mean by the molecule itself, make an internal coordinate system the most logical representation. In some fields the IC representation (bond length, angle between bonds, and twist angle of the bond as shown in the figure) is known as the Z-matrix or torsion angle representation. Unfortunately, continuous motions in Cartesian space often require discontinuous angular branches in internal coordinates making it relatively hard to work with force fields in the internal coordinate representation and conversely a simple displacement of an atom in Cartesian space may not be a straight line trajectory due to the prohibitions of the interconnected bonds. Thus it is very common for computational optimization programs to flip back and forth between representations during their iterations; this can dominate the calculation time of the potential itself and in long chain molecules introduce cumulative numerical inaccuracy. While all conversion algorithms produce mathematically identical results, they differ in speed and numerical accuracy.[1] Currently, the fastest and most accurate torsion to Cartesian conversion is the Natural Extension Reference Frame (NERF) method.[1]

23.3   Applications

Molecular modelling methods are now routinely used to investigate the structure, dynamics, surface properties and thermodynamics of inorganic, biological and polymeric systems. The types of biological activity that have been investigated using molecular modelling include protein folding, enzyme catalysis, protein stability, conformational changes associated with biomolecular function, and molecular recognition of proteins, DNA, and membrane complexes.

23.4   See also

- Cheminformatics

- Density functional theory programs.

- Force field implementation

- List of molecular graphics systems

- List of nucleic acid simulation software

- List of software for Monte Carlo molecular modelling

- List of software for nanostructures modelling

- List of protein structure prediction software

- Molecular design software

- Molecular graphics

- Molecular model

- Molecular modelling on GPU

- Molecule editor

- Monte Carlo method

- Quantum chemistry computer programs

- Semi-empirical quantum chemistry method

- Simulated reality

- Software for molecular mechanics modelling

- Structural bioinformatics

- Z-matrix

23.5 Notes

[1] Parsons, J., Holmes, J. B., Rojas, J. M., Tsai, J., Strauss, C. E., **Practical conversion from torsion space to cartesian space for in silico protein synthesis.** J Comput Chem 26 (2005), 1063-1068.

- M. P. Allen, D. J. Tildesley, *Computer simulation of liquids*, 1989, Oxford University Press, ISBN 0-19-855645-4.

- A. R. Leach, *Molecular Modelling: Principles and Applications*, 2001, ISBN 0-582-38210-6

- D. Frenkel, B. Smit, *Understanding Molecular Simulation: From Algorithms to Applications*, 1996, ISBN 0-12-267370-0

- D. C. Rapaport, *The Art of Molecular Dynamics Simulation*, 2004, ISBN 0-521-82568-7

- R. J. Sadus, *Molecular Simulation of Fluids: Theory, Algorithms and Object-Orientation*, 2002, ISBN 0-444-51082-6

- K.I.Ramachandran, G Deepa and Krishnan Namboori. P.K. *Computational Chemistry and Molecular Modeling Principles and Applications* 2008 ISBN 978-3-540-77302-3 Springer-Verlag GmbH

23.6 External links

- Center for Molecular Modeling at the National Institutes of Health (NIH) (U.S. Government Agency)

- Molecular Simulation, details for the Molecular Simulation journal, ISSN: 0892-7022 (print), 1029-0435 (online)

- The eCheminfo Network and Community of Practice in Informatics and Modeling

- Molecular Modelling Italian web portal

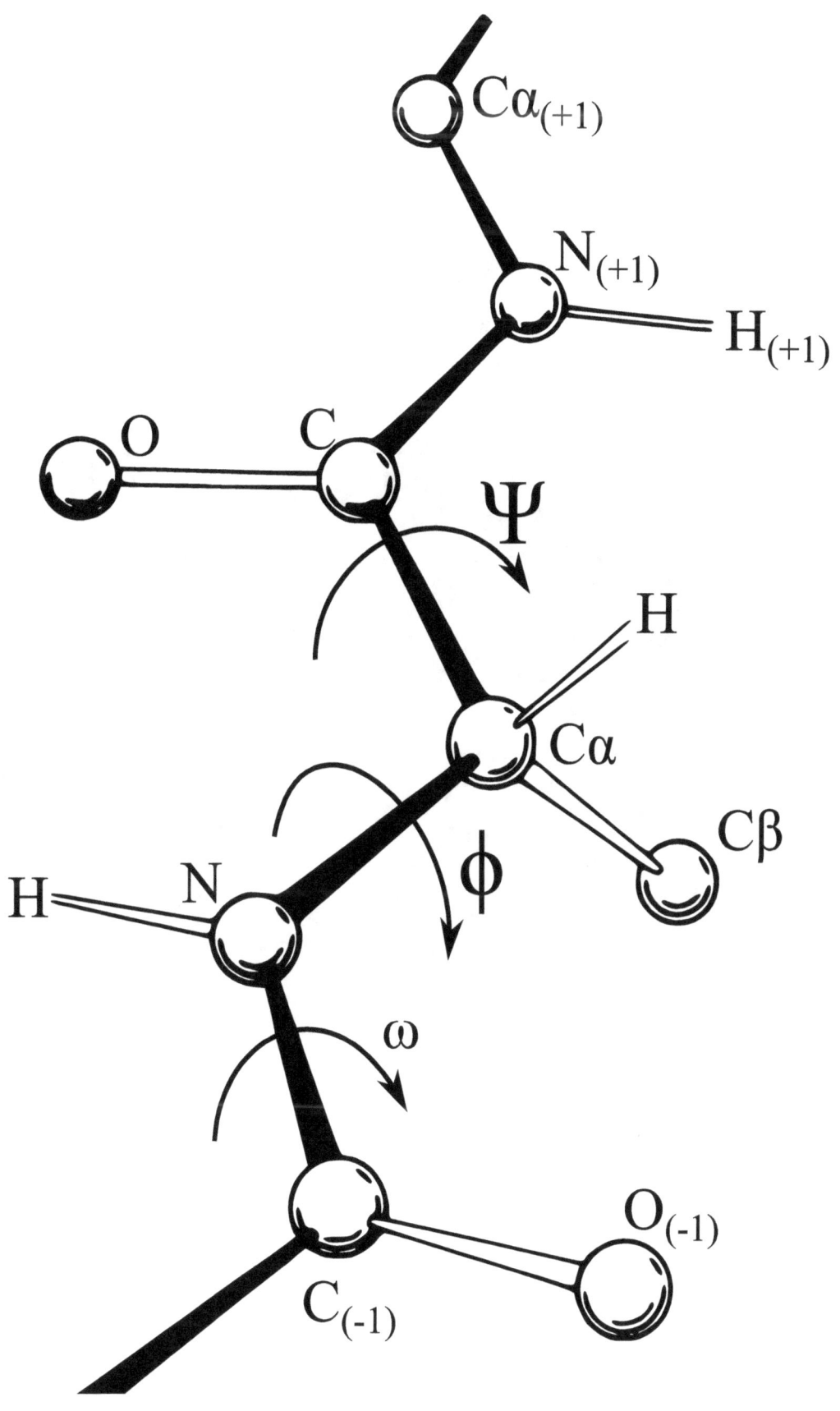

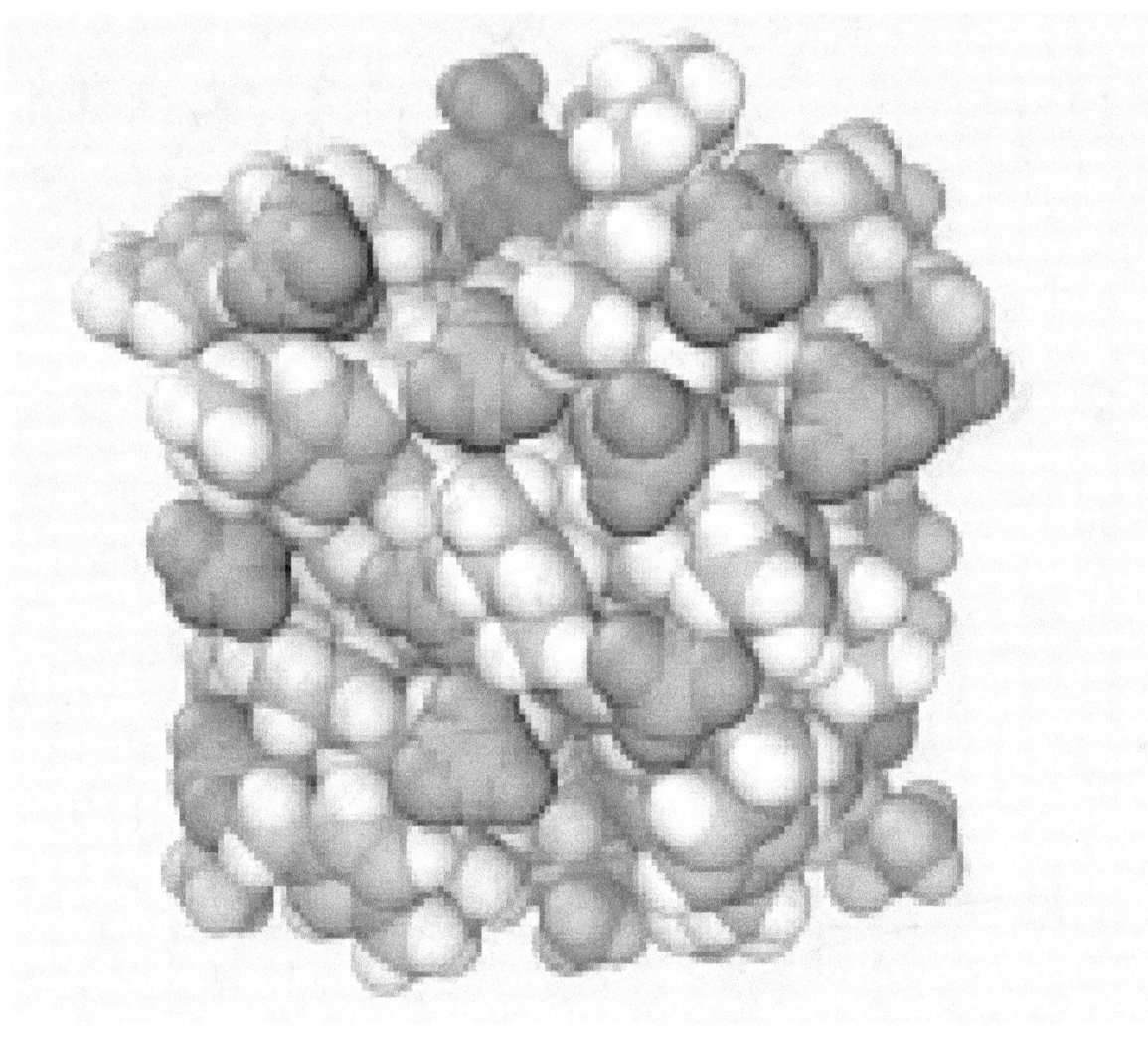

Modeling of ionic liquid

Chapter 24

Molecular orbital

See also: Molecular orbital theory

In chemistry, a **molecular orbital** (or **MO**) is a mathematical function describing the wave-like behavior of an electron in a molecule. This function can be used to calculate chemical and physical properties such as the probability of finding an electron in any specific region. The term *orbital* was introduced by Robert S. Mulliken in 1932 as an abbreviation for *one-electron orbital wave function*.[1] At an elementary level, it is used to describe the *region* of space in which the function has a significant amplitude. Molecular orbitals are usually constructed by combining atomic orbitals or hybrid orbitals from each atom of the molecule, or other molecular orbitals from groups of atoms. They can be quantitatively calculated using the Hartree–Fock or self-consistent field (SCF) methods.

24.1   Overview

A molecular orbital (MO) can be used to represent the regions in a molecule where an electron occupying that orbital is likely to be found. Molecular orbitals are obtained from the combination of atomic orbitals, which predict the location of an electron in an atom. A molecular orbital can specify the electron configuration of a molecule: the spatial distribution and energy of one (or one pair of) electron(s). Most commonly an MO is represented as a linear combination of atomic orbitals (the LCAO-MO method), especially in qualitative or very approximate usage. They are invaluable in providing a simple model of bonding in molecules, understood through molecular orbital theory. Most present-day methods in computational chemistry begin by calculating the MOs of the system. A molecular orbital describes the behavior of one electron in the electric field generated by the nuclei and some average distribution of the other electrons. In the case of two electrons occupying the same orbital, the Pauli principle demands that they have opposite spin. Necessarily this is an approximation, and highly accurate descriptions of the molecular electronic wave function do not have orbitals (see configuration interaction).

24.2   Formation of molecular orbitals

Molecular orbitals arise from allowed interactions between atomic orbitals, which are allowed if the symmetries (determined from group theory) of the atomic orbitals are compatible with each other. Efficiency of atomic orbital interactions is determined from the overlap (a measure of how well two orbitals constructively interact with one another) between two atomic orbitals, which is significant if the atomic orbitals are close in energy. Finally, the number of molecular orbitals that form must equal the number of atomic orbitals in the atoms being combined to form the molecule.

24.3 Qualitative discussion

For an imprecise, but qualitatively useful, discussion of the molecular structure, the molecular orbitals can be obtained from the "Linear combination of atomic orbitals molecular orbital method" ansatz. Here, the molecular orbitals are expressed as linear combinations of atomic orbitals.

24.3.1 Linear combinations of atomic orbitals (LCAO)

Main article: Linear combination of atomic orbitals

Molecular orbitals were first introduced by Friedrich Hund[2][3] and Robert S. Mulliken[4][5] in 1927 and 1928.[6][7] The linear combination of atomic orbitals or "LCAO" approximation for molecular orbitals was introduced in 1929 by Sir John Lennard-Jones.[8] His ground-breaking paper showed how to derive the electronic structure of the fluorine and oxygen molecules from quantum principles. This qualitative approach to molecular orbital theory is part of the start of modern quantum chemistry. Linear combinations of atomic orbitals (LCAO) can be used to estimate the molecular orbitals that are formed upon bonding between the molecule's constituent atoms. Similar to an atomic orbital, a Schrödinger equation, which describes the behavior of an electron, can be constructed for a molecular orbital as well. Linear combinations of atomic orbitals, or the sums and differences of the atomic wavefunctions, provide approximate solutions to the Hartree–Fock equations which correspond to the independent-particle approximation of the molecular Schrödinger equation. For simple diatomic molecules, the wavefunctions obtained are represented mathematically by the equations

$$\Psi = c_a \psi_a + c_b \psi_b$$

$$\Psi^* = c_a \psi_a - c_b \psi_b$$

where Ψ and Ψ^* are the molecular wavefunctions for the bonding and antibonding molecular orbitals, respectively, ψ_a and ψ_b are the atomic wavefunctions from atoms a and b, respectively, and c_a and c_b are adjustable coefficients. These coefficients can be positive or negative, depending on the energies and symmetries of the individual atomic orbitals. As the two atoms become closer together, their atomic orbitals overlap to produce areas of high electron density, and, as a consequence, molecular orbitals are formed between the two atoms. The atoms are held together by the electrostatic attraction between the positively charged nuclei and the negatively charged electrons occupying bonding molecular orbitals.[9]

24.3.2 Bonding, antibonding, and nonbonding MOs

When atomic orbitals interact, the resulting molecular orbital can be of three types: bonding, antibonding, or nonbonding.

Bonding MOs:

- Bonding interactions between atomic orbitals are constructive (in-phase) interactions.
- Bonding MOs are lower in energy than the atomic orbitals that combine to produce them.

Antibonding MOs:

- Antibonding interactions between atomic orbitals are destructive (out-of-phase) interactions, with a nodal plane where the wavefunction of the antibonding orbital is zero between the two interacting atoms
- Antibonding MOs are higher in energy than the atomic orbitals that combine to produce them.

Nonbonding MOs:

- Nonbonding MOs are the result of no interaction between atomic orbitals because of lack of compatible symmetries.
- Nonbonding MOs will have the same energy as the atomic orbitals of one of the atoms in the molecule.

24.3.3   Sigma and pi labels for MOs

The type of interaction between atomic orbitals can be further categorized by the molecular-orbital symmetry labels σ (sigma), π (pi), δ (delta), φ (phi), γ (gamma) etc. paralleling the symmetry of the atomic orbitals s, p, d, f and g. The number of nodal planes containing the internuclear axis between the atoms concerned is zero for σ MOs, one for π, two for δ, etc.

σ symmetry

For more details on this topic, see Sigma bond.

A MO with σ symmetry results from the interaction of either two atomic s-orbitals or two atomic p_z-orbitals. An MO will have σ-symmetry if the orbital is symmetrical with respect to the axis joining the two nuclear centers, the internuclear axis. This means that rotation of the MO about the internuclear axis does not result in a phase change. A σ* orbital, sigma antibonding orbital, also maintains the same phase when rotated about the internuclear axis. The σ* orbital has a nodal plane that is between the nuclei and perpendicular to the internuclear axis.[10]

π symmetry

For more details on this topic, see Pi bond.

A MO with π symmetry results from the interaction of either two atomic p_x orbitals or p_y orbitals. An MO will have π symmetry if the orbital is asymmetrical with respect to rotation about the internuclear axis. This means that rotation of the MO about the internuclear axis will result in a phase change. There is one nodal plane containing the internuclear axis, if real orbitals are considered.

A π* orbital, pi antibonding orbital, will also produce a phase change when rotated about the internuclear axis. The π* orbital also has a second nodal plane between the nuclei.[11][12][13][14]

δ symmetry

For more details on this topic, see Delta bond.

A MO with δ symmetry results from the interaction of two atomic d_{xy} or $d_{x^2-y^2}$ orbitals. Because these molecular orbitals involve low-energy d atomic orbitals, they are seen in transition-metal complexes. A δ bonding orbital has two nodal planes containing the internuclear axis, and a δ* antibonding orbital also has a third nodal plane between the nuclei.

φ symmetry

For more details on this topic, see Phi bond.

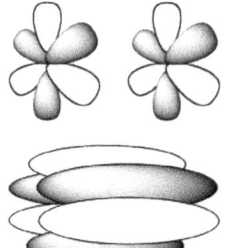

Suitably aligned f atomic orbitals overlap to form phi molecular orbital (a phi bond)

Theoretical chemists have conjectured that higher-order bonds, such as phi bonds corresponding to overlap of f atomic orbitals, are possible. There is as of 2005 only one known example of a molecule purported to contain a phi bond (a U–U bond, in the molecule U_2).[15]

24.3.4 Gerade and ungerade symmetry

For molecules that possess a center of inversion (centrosymmetric molecules) there are additional labels of symmetry that can be applied to molecular orbitals. Centrosymmetric molecules include:

- Homonuclear diatomics, X_2

- Octahedral, EX_6

- Square planar, EX_4.

Non-centrosymmetric molecules include:

- Heteronuclear diatomics, XY

- Tetrahedral, EX_4.

If inversion through the center of symmetry in a molecule results in the same phases for the molecular orbital, then the MO is said to have gerade (g) symmetry, from the German word for even. If inversion through the center of symmetry in a molecule results in a phase change for the molecular orbital, then the MO is said to have ungerade (u) symmetry, from the German word for odd. For a bonding MO with σ-symmetry, the orbital is σ_g (s' + s" is symmetric), while an antibonding MO with σ-symmetry the orbital is σ_u, because inversion of s' – s" is antisymmetric. For a bonding MO with π-symmetry the orbital is π_u because inversion through the center of symmetry for would produce a sign change (the two p atomic orbitals are in phase with each other but the two lobes have opposite signs), while an antibonding MO with π-symmetry is π_g because inversion through the center of symmetry for would not produce a sign change (the two p orbitals are antisymmetric by phase).[16]

24.3.5 MO diagrams

Main article: Molecular orbital diagram

The qualitative approach of MO analysis uses a molecular orbital diagram to visualize bonding interactions in a molecule. In this type of diagram, the molecular orbitals are represented by horizontal lines; the higher a line the higher the energy of the orbital, and degenerate orbitals are placed on the same level with a space between them. Then, the electrons to be placed in the molecular orbitals are slotted in one by one, keeping in mind the Pauli exclusion principle and Hund's rule of maximum multiplicity (only 2 electrons, having opposite spins, per orbital; place as many unpaired electrons on one energy level as possible before starting to pair them). For more complicated molecules, the wave mechanics approach loses utility in a qualitative understanding of bonding (although is still necessary for a quantitative approach). Some properties:

- A basis set of orbitals includes those atomic orbitals that are available for molecular orbital interactions, which may be bonding or antibonding

- The number of molecular orbitals is equal to the number of atomic orbitals included in the linear expansion or the basis set

- If the molecule has some symmetry, the degenerate atomic orbitals (with the same atomic energy) are grouped in linear combinations (called **symmetry-adapted atomic orbitals (SO)**), which belong to the representation of the symmetry group, so the wave functions that describe the group are known as **symmetry-adapted linear combinations (SALC)**.

- The number of molecular orbitals belonging to one group representation is equal to the number of symmetry-adapted atomic orbitals belonging to this representation

- Within a particular representation, the symmetry-adapted atomic orbitals mix more if their atomic energy levels are closer.

The general procedure for constructing a molecular orbital diagram for a reasonably simple molecule can be summarized as follows:

1. Assign a point group to the molecule.

2. Look up the shapes of the SALCs.

3. Arrange the SALCs of each molecular fragment in increasing order of energy, first noting whether they stem from s, p, or d orbitals (and put them in the order $s < p < d$), and then their number of internuclear nodes.

4. Combine SALCs of the same symmetry type from the two fragments, and from N SALCs form N molecular orbitals.

5. Estimate the relative energies of the molecular orbitals from considerations of overlap and relative energies of the parent orbitals, and draw the levels on a molecular orbital energy level diagram (showing the origin of the orbitals).

6. Confirm, correct, and revise this qualitative order by carrying out a molecular orbital calculation by using commercial software.[17]

24.3.6   Bonding in molecular orbitals

Orbital degeneracy

Main article: Degenerate orbital

Molecular orbitals are said to be degenerate if they have the same energy. For example, in the homonuclear diatomic molecules of the first ten elements, the molecular orbitals derived from the p_x and the p_y atomic orbitals result in two degenerate bonding orbitals (of low energy) and two degenerate antibonding orbitals (of high energy).[18]

Ionic bonds

Main article: Ionic bond

When the energy difference between the atomic orbitals of two atoms is quite large, one atom's orbitals contribute almost entirely to the bonding orbitals, and the others atom's orbitals contribute almost entirely to the antibonding orbitals. Thus, the situation is effectively that some electrons have been transferred from one atom to the other. This is called an (mostly) ionic bond.

Bond order

Main article: Bond order

The bond order, or number of bonds, of a molecule can be determined by combining the number of electrons in bonding and antibonding molecular orbitals. A pair of electrons in a bonding orbital creates a bond, whereas a pair of electrons in an antibonding orbital negates a bond. For example, N_2, with eight electrons in bonding orbitals and two electrons in antibonding orbitals, has a bond order of three, which constitutes a triple bond.

Bond strength is proportional to bond order—a greater amount of bonding produces a more stable bond—and bond length is inversely proportional to it—a stronger bond is shorter.

There are rare exceptions to the requirement of molecule having a positive bond order. Although Be_2 has a bond order of 0 according to MO analysis, there is experimental evidence of a highly unstable Be_2 molecule having a bond length of 245 pm and bond energy of 10 kJ/mol.[19]

HOMO and LUMO

Main article: HOMO/LUMO

The highest occupied molecular orbital and lowest unoccupied molecular orbital are often referred to as the HOMO and LUMO, respectively. The difference of the energies of the HOMO and LUMO, termed the band gap, can sometimes serve as a measure of the excitability of the molecule: The smaller the energy the more easily it will be excited.

24.4 Molecular orbital examples

24.4.1 Homonuclear diatomics

Homonuclear diatomic MOs contain equal contributions from each atomic orbital in the basis set. This is shown in the homonuclear diatomic MO diagrams for H_2, He_2, and Li_2, all of which containing symmetric orbitals.[20]

H_2

As a simple MO example consider the hydrogen molecule, H_2 (see molecular orbital diagram), with the two atoms labelled H' and H". The lowest-energy atomic orbitals, 1s' and 1s", do not transform according to the symmetries of the molecule. However, the following symmetry adapted atomic orbitals do:

The symmetric combination (called a bonding orbital) is lower in energy than the basis orbitals, and the antisymmetric combination (called an antibonding orbital) is higher. Because the H_2 molecule has two electrons, they can both go in the bonding orbital, making the system lower in energy (and, hence, more stable) than two free hydrogen atoms. This is called a covalent bond. The bond order is equal to the number of bonding electrons minus the number of antibonding electrons, divided by 2. In this example, there are 2 electrons in the bonding orbital and none in the antibonding orbital; the bond order is 1, and there is a single bond between the two hydrogen atoms.

He_2

On the other hand, consider the hypothetical molecule of He_2 with the atoms labeled He' and He". As with H_2, the lowest-energy atomic orbitals are the 1s' and 1s", and do not transform according to the symmetries of the molecule, while the symmetry adapted atomic orbitals do. The symmetric combination—the bonding orbital—is lower in energy than the basis orbitals, and the antisymmetric combination—the antibonding orbital—is higher. Unlike H_2, with two valence electrons, He_2 has four in its neutral ground state. Two electrons fill the lower-energy bonding orbital, $\sigma_g(1s)$, while the remaining two fill the higher-energy antibonding orbital, $\sigma_u^*(1s)$. Thus, the resulting electron density around the molecule does not support the formation of a bond between the two atoms; without a stable bond holding the atoms together, molecule would not be expected to exist. Another way of looking at it is that there are two bonding electrons and two antibonding electrons; therefore, the bond order is 0 and no bond exists (the molecule has one bound state supported by the Van der Waals potential).

Li_2

Dilithium Li_2 is formed from the overlap of the 1s and 2s atomic orbitals (the basis set) of two Li atoms. Each Li atom contributes three electrons for bonding interactions, and the six electrons fill the three MOs of lowest energy, $\sigma_g(1s)$, $\sigma_u^*(1s)$, and $\sigma_g(2s)$. Using the equation for bond order, it is found that dilithium has a bond order of one, a single bond.

Noble gases

Considering a hypothetical molecule of He_2, since the basis set of atomic orbitals is the same as in the case of H_2, we find that both the bonding and antibonding orbitals are filled, so there is no energy advantage to the pair. HeH would have a slight energy advantage, but not as much as $H_2 + 2$ He, so the molecule is very unstable and exists only briefly before decomposing into hydrogen and helium. In general, we find that atoms such as He that have full energy shells rarely bond with other atoms. Except for short-lived Van der Waals complexes, there are very few noble gas compounds known.

24.4.2 Heteronuclear diatomics

While MOs for homonuclear diatomic molecules contain equal contributions from each interacting atomic orbital, MOs for heteronuclear diatomics contain different atomic orbital contributions. Orbital interactions to produce bonding or antibonding orbitals in heteronuclear diatomics occur if there is sufficient overlap between atomic orbitals as determined by their symmetries and similarity in orbital energies.

HF

In hydrogen fluoride HF overlap between the H 1s and F 2s orbitals is allowed by symmetry but the difference in energy between the two atomic orbitals prevents them from interacting to create a molecular orbital. Overlap between the H 1s and F $2p_z$ orbitals is also symmetry allowed, and these two atomic orbitals have a small energy separation. Thus, they interact, leading to creation of σ and σ* MOs and a molecule with a bond order of 1. Since HF is a non-centrosymmetric molecule, the symmetry labels g and u do not apply to its molecular orbitals.[21]

24.5 Quantitative approach

To obtain quantitative values for the molecular energy levels, one needs to have molecular orbitals that are such that the configuration interaction (CI) expansion converges fast towards the full CI limit. The most common method to obtain such functions is the Hartree–Fock method, which expresses the molecular orbitals as eigenfunctions of the Fock operator. One usually solves this problem by expanding the molecular orbitals as linear combinations of Gaussian functions centered on the atomic nuclei (see linear combination of atomic orbitals and basis set (chemistry)). The equation for the coefficients of these linear combinations is a generalized eigenvalue equation known as the Roothaan equations, which are in fact a particular representation of the Hartree–Fock equation. There are a number of programs in which quantum chemical calculations of MOs can be performed, including Spartan and HyperChem.

Simple accounts often suggest that experimental molecular orbital energies can be obtained by the methods of ultra-violet photoelectron spectroscopy for valence orbitals and X-ray photoelectron spectroscopy for core orbitals. This, however, is incorrect as these experiments measure the ionization energy, the difference in energy between the molecule and one of the ions resulting from the removal of one electron. Ionization energies are linked approximately to orbital energies by Koopmans' theorem. While the agreement between these two values can be close for some molecules, it can be very poor in other cases.

24.6 References

[1] Mulliken, Robert S. (July 1932). "Electronic Structures of Polyatomic Molecules and Valence. II. General Considerations". *Physical Review* **41** (1): 49–71. Bibcode:1932PhRv...41...49M. doi:10.1103/PhysRev.41.49.

[2] F. Hund, "Zur Deutung einiger Erscheinungen in den Molekelspektren" [On the interpretation of some phenomena in molecular spectra] *Zeitschrift für Physik*, vol. 36, pages 657-674 (1926).

[3] F. Hund, "Zur Deutung der Molekelspektren", *Zeitschrift für Physik*, Part I, vol. 40, pages 742-764 (1927); Part II, vol. 42, pages 93–120 (1927); Part III, vol. 43, pages 805-826 (1927); Part IV, vol. 51, pages 759-795 (1928); Part V, vol. 63, pages 719-751 (1930).

[4] R. S. Mulliken, "Electronic states. IV. Hund's theory; second positive nitrogen and Swan bands; alternate intensities", *Physical Review*, vol. 29, pages 637–649 (1927).

[5] R. S. Mulliken, "The assignment of quantum numbers for electrons in molecules", *Physical Review*, vol. 32, pages 186–222 (1928).

[6] Friedrich Hund and Chemistry, Werner Kutzelnigg, on the occasion of Hund's 100th birthday, Angewandte Chemie International Edition, 35, 573–586, (1996)

[7] Robert S. Mulliken's Nobel Lecture, Science, 157, no. 3785, 13 - 24, (1967). Available on-line at: Nobelprize.org .

[8] Sir John Lennard-Jones, "The electronic structure of some diatomic molecules", *Transactions of the Faraday Society*, vol. 25, pages 668-686 (1929).

[9] Gary L. Miessler; Donald A. Tarr. Inorganic Chemistry. Pearson Prentice Hall, 3rd ed., 2004.

[10] Catherine E. Housecroft, Alan G, Sharpe, Inorganic Chemistry, Pearson Prentice Hall; 2nd Edition, 2005, p. 29-33.

[11] Catherine E. Housecroft, Alan G, Sharpe, *Inorganic Chemistry*, Pearson Prentice Hall; 2nd Edition, 2005, p. 29-33.

[12] Peter Atkins; Julio De Paula. *Atkins' Physical Chemistry*. Oxford University Press, 8th ed., 2006.

[13] Yves Jean; François Volatron. *An Introduction to Molecular Orbitals*. Oxford University Press, 1993.

[14] Michael Munowitz, *Principles of Chemistry*, Norton & Company, 2000, p. 229-233.

[15] Gagliardi, Laura; Roos, Björn O. (2005). "Quantum chemical calculations show that the uranium molecule U2 has a quintuple bond". *Nature* **433**: 848–851. Bibcode:2005Natur.433..848G. doi:10.1038/nature03249.

[16] Catherine E. Housecroft, Alan G, Sharpe, Inorganic Chemistry, Pearson Prentice Hall; 2nd Edition, 2005, p. 29-33.

[17] Atkins., Peter ... [et al]. (2006). *Inorganic chemistry* (4. ed.). New York: W.H. Freeman. p. 208. ISBN 978-0-7167-4878-6. Check date values in: |accessdate= (help);

[18] Gary L. Miessler; Donald A. Tarr. Inorganic Chemistry. Pearson Prentice Hall, 3rd ed., 2004.

[19] Catherine E. Housecroft, Alan G, Sharpe, Inorganic Chemistry, Pearson Prentice Hall; 2nd Edition, 2005, p. 29-33.

[20] Catherine E. Housecroft, Alan G, Sharpe, Inorganic Chemistry, Pearson Prentice Hall; 2nd Edition, 2005, p. 29-33.

[21] Catherine E. Housecroft, Alan G, Sharpe, Inorganic Chemistry, Pearson Prentice Hall; 2nd Edition, 2005, p. 41-43.

24.7 External links

- Java molecular orbital viewer shows orbitals of hydrogen molecular ion.

- The orbitron, a visualization of all atomic, and some molecular and hybrid orbitals

- xeo Visualizations of some atomic and molecular atoms

- OrbiMol Molecular orbital database.

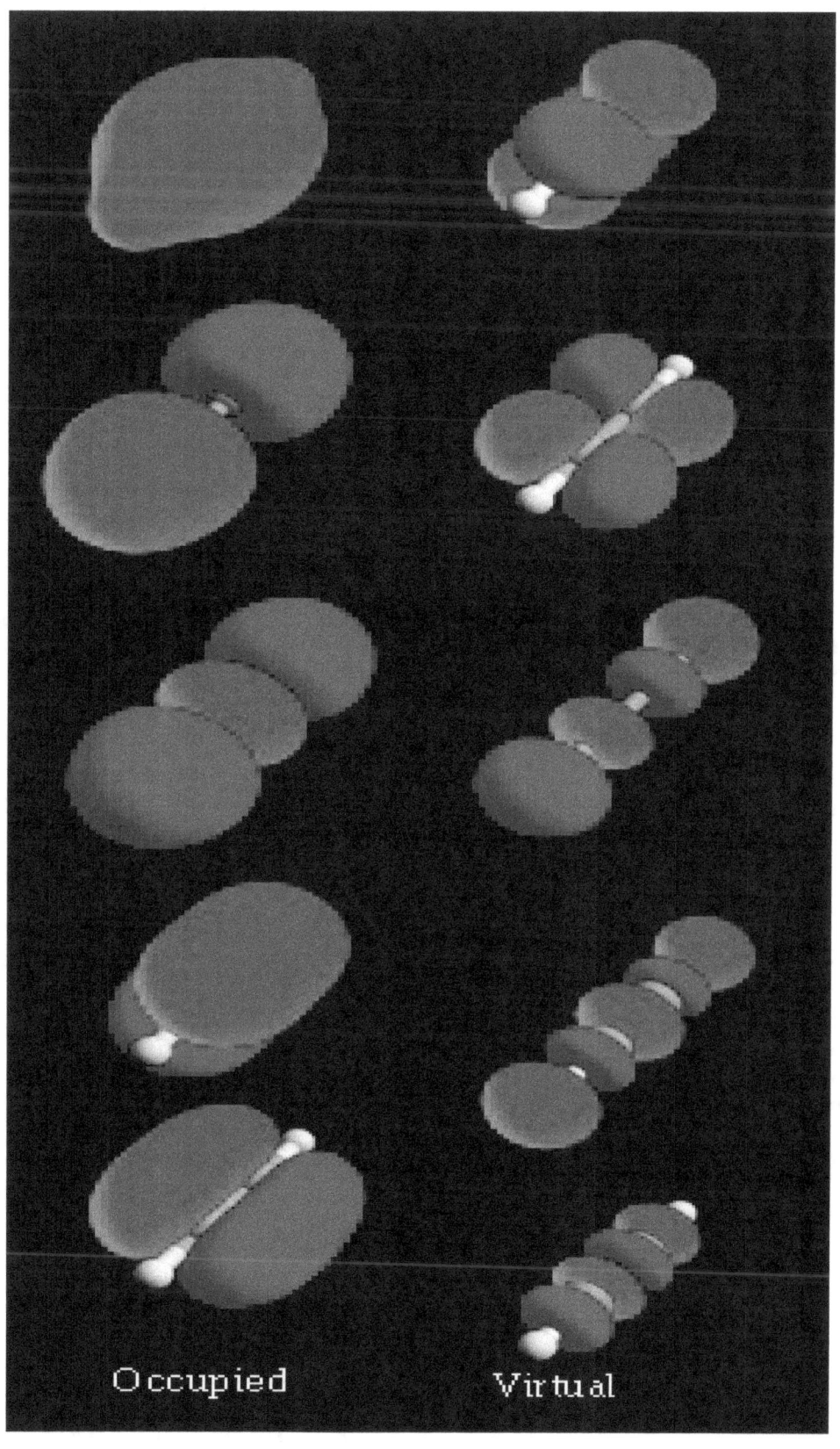

Occupied   Virtual

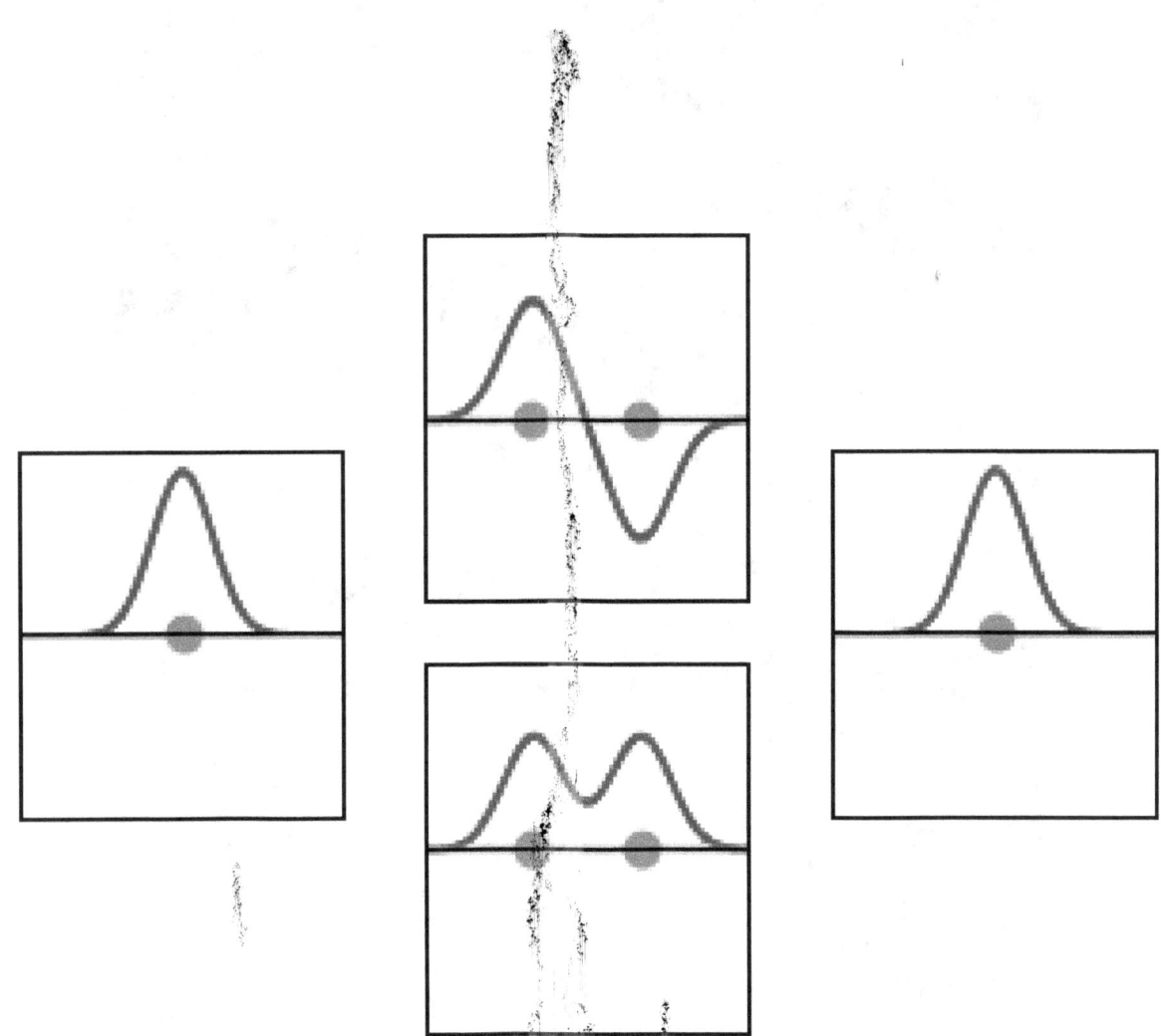

Electron wavefunctions for the 1s orbital of a lone hydrogen atom (left and right) and the corresponding bonding (bottom) and antibonding (top) molecular orbitals of the H₂ molecule. The real part of the wavefunction is the blue curve, and the imaginary part is the red curve. The red dots mark the locations of the nuclei. The electron wavefunction oscillates according to the Schrödinger wave equation, and orbitals are its standing waves. The standing wave frequency is proportional to the orbital's kinetic energy. (This plot is a one-dimensional slice through the three-dimensional system.)

Chapter 25

World Wide Molecular Matrix

The **World Wide Molecular Matrix** (WWMM) is an electronic repository for unpublished chemical data. First proposed in 2002 by Peter Murray-Rust and his colleagues in the chemistry department at the University of Cambridge in the United Kingdom, WWMM provides a free, easily searchable database for information about thousands of complicated molecules, data that would otherwise remain inaccessible to scientists.

Murray-Rust, a chemical informatics specialist, has estimated that 80% of the results produced by chemists around the world is never published in scientific journals. Most of this data is not ground-breaking, yet it could conceivably be of use to scientists doing related projects—if they could access it. The WWMM was proposed as a solution to this problem. It would house the results of experiments on over 100,000 molecules in physical chemistry, organic chemistry, biochemistry and medicinal chemistry.

In other scientific fields, the need for a similar depository to house inaccessible information could be more acute. In a presentation at the "CERN Workshop on Innovations in Scholarly Communications (OAI4)", Murray-Rust said that chemistry actually leads other fields in published data. He estimated that as much as 99% of the data in some scientific fields never reaches publication.

Although scientific in nature, the WWMM is part of the broader open archives and open source movements, pushes to make more and more information freely available to any user via the Internet or World Wide Web. In his CERN presentation, Murray-Rust stated that the WWMM was a "response to the expense of [scientific] journals," and he asked the rhetorical question, "Can we win the war to make data open, or will it be absorbed into the publishing and pseudo-publishing world?" Murray-Rust and his colleagues are also responsible for the development of the Chemical Mark-up Language (CML), a variant of XML intended for chemists.

25.1   See also

- The open archives initiative (OAI)

- The science of Informatics

- Chemical Mark-up language (CML)

25.2   External links

- The home page of Dr. Peter Murray-Rust at the University of Cambridge

- The Cambridge Center for molecular informatics

- An outline of the WWMM

- CERN Workshop on Innovations in Scholarly Communication (OAI4)

25.3   Text and image sources, contributors, and licenses

25.3.1   Text

- **Molecule** *Source:* https://en.wikipedia.org/wiki/Molecule?oldid=676534369 *Contributors:* Sodium, Tarquin, WillWare, Andre Engels, Christian List, Ben-Zin~enwiki, Lir, Patrick, Michael Hardy, Fred Bauder, Liftarn, Looxix~enwiki, Ahoerstemeier, Theresa knott, Nanobug, Jebba, Александър, Julesd, Nikai, Andres, Samuel~enwiki, Ddoherty, Coren, 4lex, Piolinfax, Saltine, Shizhao, Gakrivas, Phil Boswell, Gentgeen, Robbot, Sander123, Fredrik, Zandperl, Arkuat, Postdlf, Merovingian, Lsy098~enwiki, Academic Challenger, Caknuck, Kagredon, Hadal, Guy Peters, Marc Venot, Centrx, Giftlite, Dbenbenn, Graeme Bartlett, DocWatson42, Jmnbpt, Ævar Arnfjörð Bjarmason, Everyking, Fleminra, Curps, Bensaccount, Unconcerned, Jorge Stolfi, Bobblewik, SonicAD, Karol Langner, Icairns, Golnazfotohabadi, JohnArmagh, Deglr6328, Bluemask, Mike Rosoft, Discospinster, Hydrox, Cacycle, Vsmith, Wk muriithi, ESkog, Eric Forste, RJHall, Livajo, Art LaPella, Bobo192, Smalljim, Viriditas, Elipongo, Giraffedata, SpeedyGonsales, Kjkolb, Deryck Chan, Obradovic Goran, Jumbuck, Alansohn, Mykej, BodyTag, Walkerma, Bart133, Wtmitchell, Cburnett, John W. Kennedy, Ron Ritzman, Linas, TigerShark, Scjessey, Tygar, CiTrusD, TotoBaggins, SDC, Graham87, V8rik, Jclemens, Sjö, Drbogdan, HappyCamper, Ligulem, FlaBot, Naraht, Ian Pitchford, Admp~enwiki, RexNL, Gurch, OSt~enwiki, McDogm, Srleffler, Chobot, DVdm, Antiuser, The Rambling Man, Andel, YurikBot, Wavelength, RobotE, Brandmeister (old), Anonymous editor, Lucinos~enwiki, Ironist, Yyy, Mclparker, AeonicOmega, THB, Vb, Azazell0, Ospalh, Kyle Barbour, DeadEyeArrow, Zzuzz, Lt-wiki-bot, Ketsuekigata, Pietdesomere, Jolt76, Livitup, JoanneB, Katieh5584, Ásgeir IV.~enwiki, GrinBot~enwiki, Jade Knight, DVD R W, ChemGardener, Itub, SmackBot, Unschool, CarbonCopy, Prodego, Hydrogen Iodide, David Shear, Phaldo, Bomac, Davewild, BMunage, Lowzeewee, Edgar181, M stone, Yamaguchi⁇⁇, Gilliam, Ohnoitsjamie, Skizzik, JAn Dudík, Bduke, Miquonranger03, MalafayaBot, Droll, Moshe Constantine Hassan Al-Silverburg, Ctbolt, DHN-bot~enwiki, Sbharris, Colonies Chris, Hallenrm, Can't sleep, clown will eat me, Rrburke, Addshore, SundarBot, EdGl, DMacks, SteveLower, Sadi Carnot, Akendall, John, Scientizzle, Heimstern, Disavian, Accurizer, IronGargoyle, A. Parrot, MarkSutton, Munita Prasad, Hetar, Iridescent, Shoeofdeath, Llydawr, Sahuagin, Mulder416sBot, MightyWarrior, JForget, JohnCD, Nunquam Dormio, Exzakin, KnightLago, NickW557, FlyingToaster, Moreschi, Road Wizard, Rifleman 82, Michaelas10, Gogo Dodo, Flowerpotman, Dcandy, Tawkerbot4, Christian75, Pokeman, Omicronpersei8, Epbr123, Pajz, N5iln, Callmarcus, Headbomb, Marek69, Kendal Ozzel, GMiranda, D.H, Greg L, CTZMSC3, Escarbot, AntiVandalBot, Skynet1216, Gökhan, Res2216firestar, JAnDbot, Deflective, Husond, Captain538, MER-C, Avaya1, Andonic, PhilKnight, Magioladitis, Pedro, Bongwarrior, VoABot II, CattleGirl, CTF83!, Avicennasis, Animum, Johnbibby, Eldumpo, DerHexer, JaGa, Otvaltak, Yobol, MartinBot, Mythealias, Mermaid from the Baltic Sea, NAHID, ChemNerd, Halsall, CommonsDelinker, AlexiusHoratius, Fusion7, Nono64, PrestonH, KBlott, J.delanoy, Erajda, Abecedare, Rgoodermote, Numbo3, Hans Dunkelberg, Eliz81, Icseaturtles, James A. Stewart, Ncmvocalist, Ignatzmice, Mikael Häggström, SasukeX, Jcwf, Tomas-Bat, Bobianite, GDW13, KylieTastic, Joshua Issac, Linshukun, Useight, CardinalDan, Idioma-bot, Deor, VolkovBot, AlnoktaBOT, Atrip, Cursayer, Philip Trueman, TXiKiBoT, Cosmic Latte, Crohnie, Qxz, Someguy1221, Littlealien182, Anna Lincoln, Lradrama, KyleRGiggs, Jackfork, LeaveSleaves, Mannafredo, Katimawan2005, Thebigbendizzle, Vladsinger, Synthebot, Antixt, Jason Leach, Enviroboy, Burntsauce, Insanity Incarnate, Bobo The Ninja, AlleborgoBot, Daveberos, SieBot, Whiskey in the Jar, Restre419, Scarian, Happysailor, Flyer22, Antonio Lopez, KPH2293, Steven Crossin, Lightmouse, OKBot, Troy 07, Explicit, ClueBot, LAX, PipepBot, The Thing That Should Not Be, Starkiller88, Gawaxay, Wysprgr2005, Russellboyd, Drmies, Mild Bill Hiccup, SuperHamster, CounterVandalismBot, LizardJr8, DragonBot, Excirial, -Midorihana-, Jusdafax, Abrech, Lartoven, Rhododendrites, Tyler, Footballfan190, Versus22, SoxBot III, SteelMariner, Strangey0, Alchemist Jack, Aj00200, XLinkBot, Ivan Akira, Gnowor, Rror, Fishhead15, Locumele, Noctibus, TravisAF, ZooFari, HexaChord, Addbot, American Eagle, Some jerk on the Internet, Guoguo12, Landon1980, Otisjimmy1, Binary TSO, Looie496, Favonian, Funnybunny123456789, LinkFA-Bot, Numbo3-bot, Ehrenkater, Bluestar232, Tide rolls, Lolimgay, Legobot, Middayexpress, Luckas-bot, ZX81, 2D, Fraggle81, THEN WHO WAS PHONE?, KamikazeBot, AnakngAraw, Suntag, Eric-Wester, Jaytoneypwnz, AnomieBOT, AUG, Jim1138, IRP, Galoubet, NickK, Materialscientist, Edguy99, The High Fin Sperm Whale, Citation bot, Mcpazzo, Roboticoman, Range125, GB fan, LilHelpa, Gsmgm, Vulcan Hephaestus, Xqbot, TinucherianBot II, Sionus, JimVC3, Capricorn42, Bihco, Renaissancee, DR.smallprick, P99am, Mileycyrushater9, BoxingWear3, Jayg101, GrouchoBot, Sophus Bie, Eugene-elgato, Erik9, Magic.Wiki, LucienBOT, Tobby72, TimonyCrickets, Recognizance, Alarics, DivineAlpha, Wireless Keyboard, Shortydude3630, Citation bot 1, Rmiller8, Amplitude101, Pinethicket, Bernarddb, Tom.Reding, Akmkgp, Jschnur, RedBot, Chembrain~enwiki, Spaceflight89, Barras, Jauhienij, Juliobryant24, Jonkerz, MrX, Izzymello, David Hedlund, Tbhotch, Sideways713, DARTH SIDIOUS 2, RjwilmsiBot, TjBot, Agent Smith (The Matrix), Justinashley, Muttenjeff, DASHBot, Emaus-Bot, John of Reading, ScottyBerg, MarioFanNo1, GoingBatty, Onegumas, Winner 42, Wikipelli, K6ka, Thecheesykid, Werieth, 15turnsm, ZéroBot, CanonLawJunkie, Everard Proudfoot, Kingkamata, L Kensington, Donner60, Negovori, DASHBotAV, Mattdj2, Jimmywolfe100, E. Fokker, ClueBot NG, CocuBot, Corusant, O.Koslowski, Widr, Spannerjam, Antiqueight, Diyar se, Helpful Pixie Bot, Art and Muscle, Bibcode Bot, BG19bot, JacobTrue, Mark Arsten, Joydeep, 15chongw1, Youfailalot, NotWith, YVSREDDY, MuAlphaTheta, Rob Hurt, Oogashocka, Ducknish, JYBot, Smartypants436, Lugia2453, James Ayling, Frosty, Scooters are gay, Sammyreedy, Kevin12xd, 069952497a, Reatlas, Darth Sitges, Elie.nasrallah, Maonato, CopyEditor998, Ophara~enwiki, Briitters, Triolysat, Jarrod127, Vesperthevoid, Joystickpenguin97, Monkbot, Aamupala86, Paparoach1301, Rgesses, Hailey Girges, BetaFrisco, TROLLCOPTER, KasparBot, Mayah011213, Machkata, Sovereign Sentinel and Anonymous: 728

- **Molecular mass** *Source:* https://en.wikipedia.org/wiki/Molecular_mass?oldid=674396786 *Contributors:* AxelBoldt, Marj Tiefert, Andre Engels, Heron, Patrick, Michael Hardy, Karada, Wik, Sabbut, Xyb, Ojigiri~enwiki, Hadal, Giftlite, Bensaccount, Unconcerned, Eequor, Geni, Icairns, Sam Hocevar, Yxk~enwiki, Canterbury Tail, Discospinster, Rich Farmbrough, Cacycle, Slipstream, DcoetzeeBot~enwiki, Swid, Brian0918, Kwamikagami, Bobo192, Smalljim, Func, Danski14, Msh210, BDD, Gene Nygaard, Badgermont, MONGO, V8rik, DePiep, Margosbot~enwiki, RexNL, Chobot, Dj Capricorn, RussBot, Giro720, FF2010, JuJube, CWenger, Tim R, Itub, SmackBot, FocalPoint, Nathaniel, Slashme, KnowledgeOfSelf, Kintetsubuffalo, Skizzik, Master Jay, Papa November, SchfiftyThree, OOSCARR, Bonaparte, Brinerustle, A. B., Hgrosser, Flyguy649, DinosaursLoveExistence, Dream out loud, Stefano85, Reptile209, Nvw, JorisvS, Peterlewis, A. Parrot, Slakr, Beetstra, Griffles, Rory O'Kane, Iridescent, Lvzon, Martin Kozák, Majora4, Shrimp wong, JForget, Vaughan Pratt, Neelix, Mondorescue, Nick Y., Rifleman 82, Lugnuts, Christian75, RolfSander, Groterjan, TimVickers, WolfmanSF, Bongwarrior, VoABot II, Coleslaw75, R'n'B, Gkc, Athaenara, Cpiral, Japaulo, Idioma-bot, VolkovBot, AlnoktaBOT, SuburbanDK, TXiKiBoT, Davehi1, Vipinhari, BotKung, Synthebot, Petergans, LordofPens, SieBot, Keilana, The Evil Spartan, Yerpo, Oxymoron83, Hobartimus, Correogsk, ClueBot, Arakunem, Yamakiri, Estevoaei, Dominik-jan, DumZiBoT, Dthomsen8, Addbot, Some jerk on the Internet, Guoguo12, Fluffernutter, NjardarBot, Download, LaaknorBot, SamatBot, West.andrew.g, Dr Zimbu, Apteva, Zorrobot, Yobol, Babban12, عالم بوبحم, AnomieBOT, Materialscientist, Citation bot, Histor2008, Capricorn42, Termininja, Kevinwburns, Democat, Saehrimnir, Pinethicket, Tanweer Morshed, Bworking, What do i want to..., Jauhienij, ActiveExpression, Clarkcj12, Marie Poise, EmausBot, GoingBatty, Clostron, Tommy2010, Dcirovic, Sjoosse, Hhhippo, Aeonx, Donner60, Sven Manguard, ClueBot NG, Satellizer, Bped1985, MerlIwBot, Jblazz21, M0rphzone, Protein Chemist, Robyn2000, Tidecrusher, The Illusive Man, Dylan57474, Reatlas, J.meija, KasparBot and Anonymous: 196

- **Atomic mass unit** *Source:* https://en.wikipedia.org/wiki/Atomic_mass_unit?oldid=677422102 *Contributors:* AxelBoldt, Mav, Bryan Derksen, Tarquin, Andre Engels, Peterlin~enwiki, Heron, Patrick, Palnatoke, Lexor, Wapcaplet, Habj, Andres, Samw, The Anomebot, Jeepien, Xyb, Indefatigable, Gentgeen, Robbot, Arkuat, Ojigiri~enwiki, Wikibot, Enochlau, Giftlite, DocWatson42, Fudoreaper, TimeLord mbw, Bobblewik, Wmahan, OldakQuill, Superborsuk, Aulis Eskola, Rdsmith4, Icairns, Karl-Henner, Guanabot, Cacycle, Vsmith, DcoetzeeBot~enwiki, Bender235, Brian0918, Joanjoc~enwiki, Hurricane111, Mike Schwartz, Dungodung, Nsaa, Nik42, Arthena, Orelstrigo, Sobolewski, Wtmitchell, Dirac1933, Gene Nygaard, MIT Trekkie, LOL, MONGO, Julo, Petwil, Graham87, Rsg, Jobnikon, Zeroparallax, Rjwilmsi, Syndicate, Alban, Boccobrock, FlaBot, Gringo300, ChongDae, BitterMan, Physchim62, Chobot, The Rambling Man, YurikBot, Borgx, Jimp, DanMS, Eleassar, NawlinWiki, NickBush24, Dhollm, Mgwalker, E2mb0t~enwiki, Kkmurray, Fram, Sbyrnes321, 🉠🉠🉠 robot, ChemGardener, Itub, BonsaiViking, Brammers, H2eddsf3, Ayjaysee, Slashme, Hydrogen Iodide, Bomac, Jab843, Gilliam, Hmains, Ppntori, Bluebot, MalafayaBot, Nbarth, Epastore, Sbharris, Hgrosser, Rogermw, Bsodmike, Neo139, DinosaursLoveExistence, Hammer1980, PhiJ, SashatoBot, Doug Bell, Heimstern, Loadmaster, Xiaphias, Tuspm, BranStark, Wwallacee, Frank Lofaro Jr., CmdrObot, Kehrli, MaxEnt, Nick Y., Zginder, Christian75, Thijs!bot, Headbomb, Greg L, AntiVandalBot, Seaphoto, QuiteUnusual, DarkAudit, Gökhan, JAnDbot, Mwarren us, Daisystanton, Allstarecho, User A1, Ashishbhatnagar72, NikNaks, J.delanoy, Melamed katz, P.wormer, Nwbeeson, Kraftlos, Potatoswatter, Fylwind, Elisevil, VolkovBot, CWii, Thedjatclubrock, G3N3515, Haade, Agha Nader, Hqb, LeaveSleaves, Psyche825, Miwanya, Spinningspark, Temporaluser, Cessilie Binnings, Quantpole, JDHeinzmann, SieBot, Dawn Bard, Dzexon, Likebox, Hxhbot, Antonio Lopez, Jaccos, ClueBot, Ronkenator, Jakarr, Gsparky, Sublime5891, Tvrtko26~enwiki, Versus22, Porchcorpter, DumZiBoT, Mark Chung, Rror, Dthomsen8, SkyLined, Hess88, Algebran, Addbot, Mortense, EdgeNavidad, Non-dropframe, Vchorozopoulos, CanadianLinuxUser, Caturdayz, CarsracBot, Lightbot, Luckas Blade, Luckas-bot, Yobot, Amirobot, KamikazeBot, AnomieBOT, Paulthomas2, DemocraticLuntz, Rubinbot, Keithbob, AdjustShift, Materialscientist, Eumolpo, Xqbot, Plumpurple, Lmitchell6, Dalton1980, Br77rino, Omnipaedista, A. di M., FrescoBot, Dger, Realistix, DivineAlpha, ClickRick, Pinethicket, I dream of horses, Claudiodib, Martinvl, Alltat, Skulldyvan, Lotje, 2en, Phïï, Inferior Olive, Anczarnecki, DARTH SIDIOUS 2, EmausBot, Super48paul, Racerx11, Midnightvulture, Hhhippo, JSquish, Quondum, Flowright138, Donner60, TYelliot, Petrb, ClueBot NG, SpikeTorontoRCP, Rarity, Helpful Pixie Bot, ChrisfromStroud, Bibcode Bot, Mark Arsten, Rm1271, Hehehe12, Acratta, BattyBot, YFdyh-bot, Garamond Lethe, Mogism, Clawer67, 93, LuigiSonic57, Reatlas, Namige, SCoyWP, Marchino61, Coreyemotela, Monkbot, Chaser1234567890, Secretkeeper12, Chiranjeev1234, Sarr Cat, Buzzzzzzz and Anonymous: 248

- **History of molecular theory** *Source:* https://en.wikipedia.org/wiki/History_of_molecular_theory?oldid=674353283 *Contributors:* JWSchmidt, Stone, Graeme Bartlett, Fjarlq, ELApro, Grstain, Discospinster, Rich Farmbrough, Vsmith, Semifamous, Bletch, Ynhockey, Karnesky, Kurzon, V8rik, BD2412, Rjwilmsi, Nneonneo, Gringer, Naraht, RussBot, Okedem, Kkmurray, Itub, SmackBot, M stone, Hmains, Bduke, OrphanBot, Sadi Carnot, Rodri316, Gobonobo, Simon12, Trialsanderrors, The.Q, John courtneidge, Christian75, JamesAM, Headbomb, Milton Stanley, MER-C, Magioladitis, Gamkiller, JCraw, Nono64, DrKiernan, Deirlay, LordAnubisBOT, M-le-mot-dit, Benio76, Lamro, AdrianaW, Ajrocke, K. Aainsqatsi, Lucasbfrbot, Trang Oul, Amanafu, GregVolk, 718 Bot, Ktr101, Addbot, C6541, DOI bot, Yobot, Citation bot, Spellage, Voxii, LucienBOT, Tom.Reding, Trappist the monk, Olegrog, Cogiati, Makecat, RaptureBot, ClueBot NG, Helpful Pixie Bot, Bibcode Bot, Vokesk, Acratta, Khazar2, Monkbot, Sofia Koutsouveli and Anonymous: 32

- **Empirical formula** *Source:* https://en.wikipedia.org/wiki/Empirical_formula?oldid=664125234 *Contributors:* MartinHarper, Docu, Jebba, Gentgeen, Robbot, Gandalf61, Yosri, Bensaccount, Dmmaus, Brockert, Vanished user 1234567890, The MoUsY spell-checker, Karol Langner, Ukexpat, Avihu, Mike Rosoft, Discospinster, Vsmith, Nvj, Arthur Holland, Srbauer, Adambro, ACW, Benjah-bmm27, Graham87, The wub, Bhadani, Cyko149, Bornhj, YurikBot, Hede2000, Hellbus, Timloh, F. Cosoleto, Luk, ChemGardener, Itub, SmackBot, Hydrogen Iodide, Edgar181, Cool3, Para, Chlewbot, Thrane, BostonMA, Mwtoews, Bejnar, Kukini, Waggers, IvanLanin, RekishiEJ, Astirmays, FlyingToaster, Bvcrist, Mr Gronk, Epbr123, Danielle dk, RolfSander, Storkk, JAnDbot, Couchpotato99, Magioladitis, VoABot II, 28421u2232nfenfcenc, Schmloof, DoSiDo, Reguiieee, J.delanoy, Squeezeweasel, McSly, FJPB, Daimore, Sethant, AlnoktaBOT, RedAndr, Staka, Falcon8765, Enviroboy, Nagy, NHRHS2010, Chem tom, Khaister, Keilana, Radon210, Oiws, Oxymoron83, Stephen Shaw, The Stickler, Dabomb87, FlamingSilmaril, ClueBot, The Thing That Should Not Be, TheSmuel, Thegeneralguy, Puppy8800, Noosentaal, Joe N, DumZiBoT, Chickenmanishot, BodhisattvaBot, Jovianeye, Addbot, Dawynn, Wickey-nl, Xassassynx, Download, AgadaUrbanit, Ehrenkater, Lightbot, Quantumobserver, TheSuave, Yobot, Andreasmperu, Shootbamboo, Piano non troppo, Materialscientist, The High Fin Sperm Whale, Xqbot, Capricorn42, Spancek, Mikejens, Maddie!, GrouchoBot, قلی زادگان, Hamzah.burney, Goofster1020, Recognizance, Pinethicket, Oiseau Furtif, Merlion444, Gamewizard71, TobeBot, Defender of torch, MFHEagle123, EmausBot, Syncategoremata, Hhhippo, JSquish, Michel Awkal, Tolly4bolly, BioPupil, Xanchester, ClueBot NG, Gareth Griffith-Jones, MelbourneStar, Chester Markel, Movses-bot, Alaba2812, Tanzeela abro, Titodutta, Krenair, ISTB351, IvanVoz, EmadIV, Snow Blizzard, Matthew1585, Wherethetacos, ChrisGualtieri, JYBot, Hridith Sudev Nambiar, Yogsthefunkygirl, Harishvijayfan, CationConfiguration, Shashank5995, Asemitruck, Funbi, Ronit.Dey.357, Mandemgypsy, Chem566 and Anonymous: 185

- **Chemical substance** *Source:* https://en.wikipedia.org/wiki/Chemical_substance?oldid=677794938 *Contributors:* Marj Tiefert, Bryan Derksen, Tarquin, LA2, Enchanter, Montrealais, Patrick, Michael Hardy, Tompagenet, Mkweise, Ahoerstemeier, Notheruser, AugPi, Andres, Media lib, Ham sandwich, Tpbradbury, Geraki, Robbot, Academic Challenger, 88888888, Hadal, UtherSRG, Centrx, Giftlite, Sampo, Iridium77, Everyking, Bensaccount, Unconcerned, Eequor, Kandar, Andycjp, Alexf, Antandrus, DragonflySixtyseven, Grunners, Bluemask, Discospinster, Cacycle, Vsmith, CDN99, Bobo192, Guiltyspark, Nsaa, Alansohn, Gary, Interiot, Lectonar, Walkerma, Hohum, Snowolf, RainbowOfLight, Wimvandorst, Vuo, Gene Nygaard, Richwales, Commander Keane, Damicatz, Macaddct1984, Kralizec!, MrSomeone, Graham87, V8rik, Jake Wartenberg, SMC, Vegaswikian, Durin, DoubleBlue, Yamamoto Ichiro, Ewlyahoocom, Gurch, Mark J, Physchim62, DVdm, Gwernol, YurikBot, Wavelength, Sceptre, Flameviper, Reo On, SpuriousQ, Gaius Cornelius, Pseudomonas, NawlinWiki, Syrthiss, DeadEyeArrow, Wknight94, Silverchemist, Light current, Nikkimaria, Closedmouth, Reyk, Jolt76, DaltinWentsworth, CIreland, Kf4bdy, Itub, SmackBot, Slashme, Pavlovič, Kimon, AnonUser, Bomac, Edgar181, Gilliam, Skizzik, Chris the speller, Kurykh, Persian Poet Gal, Bduke, Fuzzform, EncMstr, SchfiftyThree, RayAYang, Achristl, FordPrefect42, Sbharris, Colonies Chris, Hallenrm, Brinerustle, Can't sleep, clown will eat me, Frap, Onorem, Yidisheryid, Addshore, PsychoCola, Zophar1, COMPFUNK2, Springnuts, SashatoBot, Accurizer, Minna Sora no Shita, IronGargoyle, Ckatz, Beetstra, Meco, Mets501, Iridescent, Wjejskenewr, IvanLanin, DavidOaks, Pudeo, Ale jrb, BoH, Fried Gold, Harej bot, Neelix, Pro bug catcher, JJC1138, Corpx, Christian75, Garik, Calvero JP, Epbr123, Headbomb, Marek69, FreeKresge,

Dawnseeker2000, CTZMSC3, AntiVandalBot, Widefox, Seaphoto, Prolog, Blu3d, Myanw, JAnDbot, Plantsurfer, Kerotan, Penubag, Parsecboy, VoABot II, JamesBWatson, Latifshaikh20, Daarznieks, DerHexer, Wdflake, Seba5618, SquidSK, Mithras6, Phoogenb, Hdt83, MartinBot, Bbi5291, ChemNerd, CommonsDelinker, Paranomia, J.delanoy, Pharaoh of the Wizards, Ginsengbomb, McSly, HiLo48, NewEnglandYankee, EyebrowOnVacation, KylieTastic, Joshua Issac, Fragbase, WJBscribe, Treisijs, FanCon, Neod4000, CardinalDan, Idioma-bot, Lights, VolkovBot, CWii, Indubitably, JohnBlackburne, AlnoktaBOT, Omar737, Crohnie, Anna Lincoln, Qyt, Seraphim, Leafyplant, Goodlilstevie, Synthebot, Dozer199, Ratear, Enviroboy, Cvf-ps, Kbrose, SieBot, Scarian, WereSpielChequers, Caltas, Yintan, Maha101, Keilana, Bob98133, Oxymoron83, Nuttycoconut, Jdaloner, Alex.muller, LonelyMarble, Paulinho28, Tony Webster, JL-Bot, Laburke, Atif.t2, Tanvir Ahmmed, ClueBot, Binksternet, Fyyer, Trevor242, Uncle Milty, SuperHamster, CounterVandalismBot, Graysen98, Soccerfreakbry, Excirial, Jusdafax, PixelBot, Abrech, Ykhwong, Coinmanj, NuclearWarfare, Arjayay, Hullcrushdepth, Kemo1993, Thingg, Horselover Frost, Bobhasissues, Saenzc, Vanished user uih38riiw4hjlsd, DumZiBoT, KingsOfHearts, Dexter siu, Zhile, Ismx24, Mkcas, Skarebo, SilvonenBot, ElMeBot, Osarius, Addbot, Metagraph, Fieldday-sunday, GD 6041, Leszek Jańczuk, LaaknorBot, CarsracBot, Glane23, AndersBot, Quercus solaris, 5 albert square, Jasoneric, Nnedass, Tide rolls, WikiDreamer Bot, Slgcat, Alfie66, Jack who built the house, Legobot II, Fryderyk, محبوب عالم, AnomieBOT, Shootbamboo, Jim1138, IRP, Dwayne, Odlaw, Materialscientist, Spy5295, The High Fin Sperm Whale, Virionspiral, Auther175, Jameson812, Rehansaccount, LilHelpa, Princesstwilighter, Xqbot, Cureden, Capricorn42, Frink99887, NFD9001, Ilyilyily, C+C, Rsmn, Lo28, Earlypsychosis, RibotBOT, میرزاحسینی حامد میرزاحسینی, Karlluo, Doulos Christos, Sophus Bie, Lordvisucius, JayJay, Pepper, Orhanghazi, Finalius, Ιωάννης Καραμήτρος, Drew R. Smith, Krish Dulal, ⿻⿻⿻⿻⿻⿻, Pinethicket, I dream of horses, A8UDI, V.narsikar, RedBot, Eagle-0, Mnjh, Lrobb95, ActivExpression, ConsumerEducation, Vrenator, Zvn, Seahorseruler, Merlinsorca, Diannaa, Suffusion of Yellow, Wikiman1222, Tbhotch, DARTH SIDIOUS 2, MMS2013, Galloping Moses, Wintonian, EmausBot, Randy madman on roids, Pavlo Chemist, RenamedUser01302013, Tommy2010, ZéroBot, Fæ, Deeas, Jande417, Chemicalinterest, Jdwwilson, Wayne Slam, Ocaasi, Wagino 20100516, Aleksander Sestak, Donner60, Orange Suede Sofa, Theislikerice, Krazie808, Neil P. Quinn, Rocketrod1960, Socialservice, ClueBot NG, AluminumFear, MelbourneStar, Pika32141, Widr, Antiqueight, Wegothim, BG19bot, Kwells1989, Swagggurl2, MusikAnimal, Frze, AvocatoBot, Mark Arsten, IraChesterfield, ZRRSZR, Snow Blizzard, Bleep234, Shubh12, GENIUS28, IsraphelMac, Webclient101, Reatlas, Leprof 7272, Riddler23tron, PlanetEditor, Www.winner,com, Uberaccount, Gayu3, Haminoon, Whitelawkirk, Susan.grayeff, WZ-121, Suelru, Ananagram, Wiki3457, Littlemonkey14, BaconMuncher57, TheQ Editor, Dfreshlut, Jimmyblocks, Last Man Never, Yakuza jackman, Argi 12345, Dog Poopie, KasparBot, Adarshjchandran, Eballard1214, LandedEagle and Anonymous: 533

- **Chemical element** *Source:* https://en.wikipedia.org/wiki/Chemical_element?oldid=677625265 *Contributors:* AxelBoldt, Vicki Rosenzweig, Mav, Bryan Derksen, The Anome, Tarquin, Css, Andre Engels, Youssefsan, Rmhermen, William Avery, Heron, Mercury610, Stevertigo, Patrick, RTC, Menchi, Ixfd64, Eric119, Minesweeper, Kosebamse, Ahoerstemeier, Suisui, Александър, Glenn, Andres, Evercat, Mxn, Schneelocke, Mulad, Reddi, Stone, Jay, Taxman, Thue, Bevo, Xevi~enwiki, Geraki, Archivist~enwiki, Donarreiskoffer, Gentgeen, Robbot, Fredrik, BitwiseMan, Yelyos, Nurg, Romanm, Arkuat, Merovingian, Pingveno, Academic Challenger, PxT, Rursus, Texture, Roscoe x, Caknuck, Wikibot, Ebeisher, Jimduck, Hexii, Centrx, Giftlite, DocWatson42, Christopher Parham, Tom harrison, HangingCurve, Xerxes314, Everyking, No Guru, Alison, Bensaccount, Rpyle731, Mboverload, R. fiend, Blankfaze, Antandrus, Beland, Karol Langner, Thincat, Icairns, Trevor MacInnis, Mike Rosoft, Sdrawkcab, HedgeHog, EugeneZelenko, Discospinster, Rich Farmbrough, FT2, Cacycle, FiP, Vsmith, Xezbeth, Nvj, Mani1, Blade Hirato~enwiki, SpookyMulder, Bender235, Andrejj, RJHall, MisterSheik, Kwamikagami, Mwanner, Laurascudder, Shanes, RoyBoy, Bookofjude, Danshil, Femto, CDN99, Bobo192, Smalljim, Cmdrjameson, SpeedyGonsales, Man vyi, Jojit fb, PeterisP, Obradovic Goran, Nsaa, Ranveig, Jumbuck, Kuratowski's Ghost, Alansohn, Hi ruwen, Loa, Paleorthid, Riana, Walkerma, Jaw959, Bantman, Sobolewski, Wtmitchell, Maxkirk1, TenOfAllTrades, LFaraone, H2g2bob, Bsadowski1, Skatebiker, Gene Nygaard, Kay Dekker, Benoni, Thryduulf, Firsfron, Alvis, SNPP, OwenX, ScottDavis, Benbest, Polyparadigm, Pol098, WadeSimMiser, Dozenist, Terence, Sengkang, CharlesC, Wayward, ⿻⿻⿻⿻⿻⿻, Mandarax, Tslocum, SqueakBox, Graham87, Ryoung122, Chun-hian, Kushboy, DePiep, Effeietsanders, Saperaud~enwiki, Rjwilmsi, Kinu, Strait, VogonFord, Tangotango, Crazynas, ScottJ, Brighterorange, ThePoorGuy, Yamamoto Ichiro, FlaBot, RobertG, Nivix, Ayla, Glenn L, Physchim62, Snailwalker, Imnotminkus, King of Hearts, Chobot, GangofOne, DVdm, Korg, NSR, Gwernol, Banaticus, Wavelength, Hawaiian717, Alchemy pete, Phantomsteve, RussBot, Ismaeelah, Limulus, Quintusdecimus, Wimt, RadioKirk, NawlinWiki, Grafen, Brythain, Peter Delmonte, E rulez, Raven4x4x, Nick C, Zwobot, Werdna, SamuelRiv, 21655, Cynicism addict, Orbis 3, NielsenGW, Peter, Johnpseudo, Katieh5584, Kungfuadam, JDspeeder1, Luk, ChemGardener, Itub, Winick88, Mexistache, Yakudza, SmackBot, Dreamer.redeemer, CarbonCopy, Hydrogen Iodide, Melchoir, C.Fred, Bomac, Jagged 85, Jrockley, Delldot, Edgar181, Half-Shadow, Alsandro, Yamaguchi⿻⿻, Gilliam, Aaron of Mpls, Skizzik, Chris the speller, Kurykh, Persian Poet Gal, MK8, Miquonranger03, MalafayaBot, CherryT~enwiki, SchfiftyThree, Bonaparte, Deli nk, Aclwon, DHN-bot~enwiki, Sbharris, Darth Panda, MaxSem, Modest Genius, Can't sleep, clown will eat me, MyNameIsVlad, Kristbg, Yidisheryid, Booshank, EvelinaB, Rrburke, Andy120290, RedHillian, SundarBot, Nakon, Lordshaun, Het, Sadi Carnot, Curly Turkey, Khazar, John, Euchiasmus, Kipala, Sir Nicholas de Mimsy-Porpington, Tony Corsini, Anoop.m, Madris, IronGargoyle, Hvn0413, Beetstra, Maksim L., Funnybunny, Iridescent, Joseph Solis in Australia, Shoeofdeath, J Di, Cbrown1023, Gil Gamesh, Civil Engineer III, Tawkerbot2, Lahiru k, JForget, Svlad Jelly, CmdrObot, Irwangatot, FunPika, Van helsing, NickW557, WeggeBot, Darren10000, Nmacu, Funnyfarmofdoom, Nilfanion, TJDay, Nick Y., Steel, Kaldosh, Rifleman 82, Gogo Dodo, Julian Mendez, Tawkerbot4, Carstensen, Christian75, DumbBOT, Obrian7, Lee, Matwilko, Mydoghasworms, Smeazel, Thijs!bot, Epbr123, Headbomb, Canada Jack, Marek69, Davidhorman, Kiran201193, ZeekyH.bomb, SusanLesch, Escarbot, Morgana The Argent, KrakatoaKatie, AntiVandalBot, Jj137, Farosdaughter, Gdo01, Istartfires, Nousakan, Res2216firestar, MER-C, Skomorokh, Plantsurfer, Hamsterlopithecus, Cynwolfe, GoodDamon, Acroterion, Moni3, Karlhahn, Bongwarrior, VoABot II, AuburnPilot, Edmund372, WhatamIdoing, Dirac66, 28421u2232nfenfcenc, Thibbs, TekNOSX, Glen, DerHexer, JaGa, T55648L, Pax:Vobiscum, TheRanger, Gwern, Rickterp, Nietzscheanlie, MartinBot, ChemNerd, Rettetast, R'n'B, AlexiusHoratius, PrestonH, Smokizzy, Tgeairn, J.delanoy, Pharaoh of the Wizards, Trusilver, Bogey97, Rhinestone K, Maurice Carbonaro, Metrax, Rod57, Chaveyd, Warut, Pcfjr9, Mufka, Tanaats, Ionescuac, Lilwik, KylieTastic, Lordaraq, Rpr117, Jamesofur, DorganBot, Treisijs, Useight, Fusion Power, Xiahou, CardinalDan, Daz643, Xenonice, Sumo su, 28bytes, VolkovBot, ABF, Jeff G., JoeDeRose, FutharkRed, Philip Trueman, TXiKiBoT, The Original Wildbear, Gary Levell, Daydreammbeliever15, Billiards, Qxz, Lradrama, Martin451, LeaveSleaves, Karlengblom, Inx272, Michelle192837, Brainmuncher, Sarc37, Wolfrock, Synthebot, Burntsauce, Riversong, 555zozo555, Onceonthisisland, AlleborgoBot, EmxBot, Glennklockwood, Demmy100, SieBot, Stever Augustus, Buccaneerande, Sonicology, PlanetStar, Winchelsea, Gerakibot, Dawn Bard, Viskonsas, Caltas, Calabraxthis, JerrySteal, Keilana, Toddst1, Tiptoety, Qst, Oda Mari, Wombatcat, Oxymoron83, Jdaloner, Lightmouse, Mjkhfg, Tombomp, Sjn28, Maelgwnbot, Wuhwuzdat, Tesi1700, Maralia, Dolphin51, Zbisasimone, Nergaal, Escape Orbit, Into The Fray, Romit3, SallyForth123, Twinsday, Elassint, ClueBot, PleasantPheasant, The Thing That Should Not Be, Rodhullandemu, VQuakr, J8079s, Boing! said Zebedee, Xenon54, Daracul, Firzen the Great, Neander7hal, Excirial, Alexbot, PrincealiG, Eeekster, Clutchmetal, ParisianBlade, Helenginn, LarryMorseDCOhio, Muro Bot, La Pianista, Calor, Thingg,

Kevin12xd, Vamsi manikanta, Valites, DudeWithAFeud, Atomdgbgfhffvgxhgdggdvhfc, Trackteur, Knowingeverything101, Saddas123sad, Ivana Pupu, Toriwood72, KasparBot and Anonymous: 549

- **Structural formula** *Source:* https://en.wikipedia.org/wiki/Structural_formula?oldid=675328890 *Contributors:* Tarquin, Jaknouse, Michael Hardy, Stone, Donarreiskoffer, Gentgeen, Bensaccount, OldakQuill, Beland, Discospinster, Cacycle, Vsmith, Mulder1982, Bobo192, Jag123, Benjah-bmm27, Walkerma, KingTT, TenOfAllTrades, Angr, Mpatel, Palica, YurikBot, Wavelength, Postglock, GeeJo, Nemu, SmackBot, Hansonrstolaf, Liaocyed, Eskimbot, Edgar181, BirdValiant, Ben.c.roberts, MalafayaBot, Sbharris, Deenoe, DJHasis, Philipwhiuk, DMacks, FrozenMan, Mets501, PSUMark2006, JForget, Kupirijo, Tawkerbot4, Christian75, Omicronpersei8, Thijs!bot, Danielle dk, WVhybrid, Anti-VandalBot, Roman à clef, Couchpotato99, Balloonguy, Dirac66, 28421u2232nfenfcenc, Allstarecho, Patstuart, Rettetast, CommonsDelinker, J.delanoy, Hans Dunkelberg, Mikael Häggström, Kloisiie, VolkovBot, Eubulides, SieBot, ClueBot, Cmj91uk, Tim32, Kurdo777, Addbot, Xp54321, Wickey-nl, Luckas-bot, Yobot, Daniele Pugliesi, LilHelpa, Brane.Blokar, Elvim, Br77rino, قلى زادگان, FrescoBot, PigFlu Oink, Armando-Martin, JSquish, Odysseus1479, Jbergste, BioPupil, ClueBot NG, Jkwchui, Widr, MerlIwBot, Juro2351, Frosty, Library computer, JPaestpreornJeolhlna, YarLucebith, Hazzalomasrico, Vasanthrojin and Anonymous: 83

- **Stereoisomerism** *Source:* https://en.wikipedia.org/wiki/Stereoisomerism?oldid=668446518 *Contributors:* Tobias Hoevekamp, Sodium, Tim Starling, Masterdlx, Diberri, PeterC, DragonflySixtyseven, Discospinster, Cacycle, Vsmith, SemperBlotto, Ceyockey, Optimusnauta, Mpatel, V8rik, FlaBot, Physchim62, YurikBot, Itub, SmackBot, Edgar181, Gilliam, Keegan, Anespa, Akriasas, Drphilharmonic, DMacks, Ohconfucius, SashatoBot, JorisvS, Beetstra, Dave Runger, Kushal one, Outriggr, Rifleman 82, D9qhd8, Steviedpeele, Aastr, Thijs!bot, Danielle dk, Davidhorman, Imoeng, Spencerw, VolkovBot, AlnoktaBOT, Flopster2, Mannafredo, Jkc0113, Undead warrior, SieBot, Faradayplank, KathrynLybarger, Michał Sobkowski, Sabri76, Irhunt, Andyp114, Myceteae, Addbot, Some jerk on the Internet, Veltraum, Yobot, Monosubst alkane, AnomieBOT, Choij, PhaseChanger, Johnny52008, Elvim, Aa77zz, Habemus, Kingpin IIT, Praveen jayaram, Markus00000, EmausBot, Vramasub, Tolly4bolly, ChuispastonBot, 28bot, Teaktl17, ClueBot NG, Mesoderm, Gauravjuvekar, ElphiBot, Nikos 1993, Hostager, PondaRox, JUNNI757, Project Osprey, Nigellwh, Iwantfreebooks, Sofia Koutsouveli and Anonymous: 103

- **Stoichiometry** *Source:* https://en.wikipedia.org/wiki/Stoichiometry?oldid=677296279 *Contributors:* Marj Tiefert, Brion VIBBER, Jan Hidders, William Avery, Maury Markowitz, Heron, Michael Hardy, Shellreef, Shimmin, Tregoweth, Weißnix, Doradus, Wetman, Johnleemk, Gentgeen, Robbot, Academic Challenger, Humus sapiens, Hadal, Giftlite, DocWatson42, Herbee, Bkonrad, Bensaccount, Gotanda, Junkyardprince, Pdefer, Icairns, Atemperman, Neutrality, Christoph Poggemann~enwiki, DanielCD, Ma'ame Michu, Vsmith, Alistair1978, Kwamikagami, Edward Z. Yang, Bobo192, Stesmo, Hagerman, Alansohn, Arthena, Diego Moya, Mbloore, BRW, RainbowOfLight, Vuo, Reinoutr, StradivariusTV, Robert K S, The Lightning Stalker, Mandarax, Graham87, BD2412, Jclemens, Mendaliv, Rjwilmsi, Vary, STarry, FlaBot, Margosbot~enwiki, Crazycomputers, Rune.welsh, Gurch, The Goog, Semi-awesome, Physchim62, Sharkface217, DVdm, Cactus.man, YurikBot, Stephenb, Megastar, Deskana, Brandon, Mysid, Kkmurray, Typer 525, Pb30, Digfarenough, Peter, Nobbysworld, Itub, SmackBot, Prodego, Istvan, David Shear, Eskimbot, TechEdge, Chris the speller, Bluebot, SchfiftyThree, Rickythesk8r, Darth Panda, SundarBot, Bolonium, Radagast83, Cybercobra, Dream out loud, Dreadstar, Drphilharmonic, The PIPE, Ashi Starshade, Pilotguy, Kukini, Akendall, Hsauro, Mbeychok, Knights who say ni, NcSchu, Rayleung2709, Mets501, Flipperinu, P199, JoeBot, Shoeofdeath, Tawkerbot2, Wolfdog, TheTito, Brownlee, Slazenger, Vanished user vjhsduheuiui4t5hjri, Master son, Odie5533, Sir Grant the Small, Viridae, Mtpaley, Omicronpersei8, Redleader666, Epbr123, 24fan24, NorwegianBlue, Mentifisto, AntiVandalBot, RapidR, Salgueiro~enwiki, Qwerty Binary, Deflective, Xoneca, VoABot II, Ambrosia-, Dekimasu, Tupeliano~enwiki, Dirac66, User A1, Adriaan, MartinBot, Mythealias, Mermaid from the Baltic Sea, TechnoFaye, Alextradewell, Wlodzimierz, J.delanoy, Star wars nerd91, Lucaswilkins, Brien Clark, Brant.merrell, Katalaveno, Stan J Klimas, Antony-22, Ross Fraser, SoCalSuperEagle, Koedinger, OliviaGuest, Teledildonix314, Philip Trueman, Axiosaurus, Brandonrush, Wolfraem, בל יבול, Staka, Insane-Contrast, Logan, Eloc Jcg, Tiddly Tom, Dawn Bard, Smilesfozwood, Troy 07, Webridge, ClueBot, NickCT, Parvazbato59, LizardJr8, Warreng9999, Auntof6, Ktr101, CohesionBot, GreenGourd, Razorflame, Dekisugi, Micvac, 1ForTheMoney, Aitias, Jeff.sloan1, Versus22, Caketastic1, Goodvac, Djh101, Bearsona, Fastily, Djyaron, Rabidclam, Ost316, NellieBly, Badgernet, Osarius, Klundarr, Addbot, Some jerk on the Internet, Tcncv, Ronhjones, EconoPhysicist, Shrey1165, Koppas, Jammon253, Tide rolls, PV=nRT, Abduallah mohammed, Yobot, Freikorp, SwisterTwister, محبوب عالم, Daniele Pugliesi, IRP, Areis1992, Histor2008, Capricorn42, Elvim, Acebulf, Nickkid5, Omnipaedista, RibotBOT, Princefuture, Sophus Bie, Shadowjams, Dougofborg, Fortdj33, Maruchan1, Doodmaker, Pinethicket, Triplestop, Sultanofhyd, Hoo man, Jschnur, December21st2012Freak, Gamewizard71, Kajervi, A p3rson, JV Smithy, Spectabilis, Minimac, Muneeb2k, Mittinatten, DARTH SIDIOUS 2, Guerillero, Mean as custard, DenaChemistry, Immunize, Vncm, IncognitoErgoSum, RenamedUser01302013, Tommy2010, Jakedog730, Jsjsjs1111, JSquish, Qniemiec, Tolly4bolly, Rocky Lorey, Scientific29, Puffin, Orange Suede Sofa, Micahd02, Liuthar, Signalizing, ClueBot NG, BubblyWantedXx, KDS444, Phalgun.lolur, Widr, Danielhumphrey, MerlIwBot, Helpful Pixie bot, JackedASS23, AsherPicklebutt, Skunkle7878, Mossy2100, Carlugo, Hghyux, Albert Maxwell, Lugia2453, JustAMuggle, Epicgenius, JellyBean4.1, Danialisboss, Brainiacal, Csutric, TreebeardTheEnt, Formula8, Ktlabe, Aaditya Upadhyay, Adwait natu, The Chemistry Bookworm, Lollipopgeek, Billy big feet, Maryaa Sillvaa and Anonymous: 412

- **Spectroscopy** *Source:* https://en.wikipedia.org/wiki/Spectroscopy?oldid=675969260 *Contributors:* Marj Tiefert, Sodium, Mav, Bryan Derksen, Andre Engels, Christian List, DrBob, Heron, Michael Hardy, Tim Starling, Pit~enwiki, Ahoerstemeier, Rob Hooft, Smack, Pizza Puzzle, Charles Matthews, Harris7, Reddi, Stone, Dysprosia, Wik, Taxman, Gentgeen, Robbot, Hankwang, Pigsonthewing, Fredrik, Arkuat, TimothyPilgrim, Giftlite, Graeme Bartlett, Snags, BenFrantzDale, Fastfission, Curps, Bensaccount, Tom-, Al-khowarizmi, Andycjp, Pcarbonn, H Padleckas, Chemstry isMyLife, Icairns, Sam Hocevar, Tsemii, Kevyn, Deglr6328, ErikvDijk, Thorwald, Vsmith, Wk muriithi, Bender235, Dkroll2, El C, RoyBoy, Femto, WiKi, Danski14, Alansohn, Atlant, SemperBlotto, Cortonin, Wtmitchell, TenOfAllTrades, Mikeo, Zereshk, Tukan~enwiki, Ken6en, Nuno Tavares, Camw, StradivariusTV, JeremyA, Mrs Trellis, Will.i.am, Pharmacomancer, Duncanssmith, Rnt20, Drbogdan, Rjwilmsi, Prgo, FlaBot, ACrush, Gurch, Srleffler, Zotel, Chobot, Hall Monitor, Adoniscik, YurikBot, Wavelength, Borgx, Charles Gaudette, Shell Kinney, Gaius Cornelius, Eleassar, Annabel, Welsh, Retired username, Fredericks, Syrthiss, BOT-Superzerocool, Bota47, Kkmurray, Peter, RG2, GrinBot~enwiki, Mejor Los Indios, SmackBot, FocalPoint, Erwinrossen, Melchoir, Edgar181, Srnec, Shovskowska, Persian Poet Gal, PDManc, MalafayaBot, Voyajer, Wen D House, Radagast83, T-borg, Hgilbert, Drphilharmonic, DMacks, Guillermo on sus ruedas, Peterlewis, IronGargoyle, Dicklyon, AmberRobot, Jaeger5432, JohnCD, WeggeBot, Korandder, Safalra, Ahmerpk111, Holdendp, Zroutik, Ignoramibus, Christian75, Thijs!bot, Epbr123, Kilva, Ishdarian, Wpostma, Headbomb, Pjvpjv, Marek69, John254, Martin Hedegaard, Aadal, Gerkleplex, AntiVandalBot, Nbkoneru, Lodder545, Prolog, Mihano, Goldenrowley, Pro crast in a tor, North Shoreman, JAnDbot, Deflective, Matthew Fennell, Ikanreed, JNW, J2thawiki, Philg88, DinoBot, R'n'B, CommonsDelinker, Ctroy36, Martyjmch, Austin512, Pyrospirit, Jcwf, Kraftlos, Williamheyn, Nicktaylor100, WinterSpw, Pdcook, Frikensmurf, Funandtrvl, Mts0405, Akhram, VolkovBot, TXiKiBoT, Oshwah, Harbir93, Qxz, Ronningt, Jameswkb, LeaveSleaves, Cremepuff222, BotKung, Sbialkow, Dfbaum, RaseaC, Spinningspark,

- **Table of permselectivity for different substances** *Source:* https://en.wikipedia.org/wiki/Table_of_permselectivity_for_different_substances?oldid=614818955 *Contributors:* Primaryspace, Mikael Häggström, Eeekster, Yobot, Macholl, Kwl-pakz and Anonymous: 1

- **List of interstellar and circumstellar molecules** *Source:* https://en.wikipedia.org/wiki/List_of_interstellar_and_circumstellar_molecules?oldid=673895697 *Contributors:* Bryan Derksen, CBDunkerson, Dbenbenn, Graeme Bartlett, Dratman, BrendanRyan, Jorge Stolfi, Foobar, Darrien, Karol Langner, Spiffy sperry, Tompw, RJHall, Huntster, Bnikolic, Pearle, Snarfevs, John Coupe, EagleFalconn, Shoefly, ThomasWinwood, Drbogdan, Rjwilmsi, Marasama, Mike s, Mike Peel, HappyCamper, Takometer, Bgwhite, Spacepotato, Deville, Reyk, Poulpy, Tropylium, SmackBot, Edgar181, Chris the speller, Bduke, Modest Genius, Chlewbot, JohnI, AstroChemist, RekishiEJ, CmdrObot, Cydebot, Astrochemist, Rifleman 82, Nikopoley, Headbomb, Pixelface, JAnDbot, Jingxin, Magioladitis, BatteryIncluded, Nono64, Leyo, James McBride, Lightmouse, Nergaal, ImageRemovalBot, NuclearWarfare, Scog, BSmith821, Ost316, Addbot, DOI bot, LinkFA-Bot, Yobot, AnomieBOT, Citation bot, Blundgr2, Hcnhplus, Citation bot 1, Jonesey95, Tom.Reding, Mikespedia, Trappist the monk, InvaderXan, Jynto, RjwilmsiBot, John of Reading, GoingBatty, TuHan-Bot, H3llBot, Whoop whoop pull up, Kikichugirl, Frietjes, Polskivinnik, Helpful Pixie Bot, Bibcode Bot, BG19bot, BattyBot, Dexbot, Ruby Murray, Eyesnore, Monkbot and Anonymous: 34

- **List of software for molecular mechanics modeling** *Source:* https://en.wikipedia.org/wiki/List_of_software_for_molecular_mechanics_modeling?oldid=668067822 *Contributors:* Michael Hardy, Thorwald, Ghutchis, Enric Naval, Mndoci, Drbreznjev, Kkmurray, Bduke, Colonies Chris, Jedgold, Baoilleach, Alma-Tadema, Leevanjackson, Hwttdz, Alaibot, Mauroesguerroto, TDF, Xebvor, Spud Gun, Mollwollfumble, Vi2, Mintz l, Ahshabazz, Kiplingw, UnitedStatesian, Michaeldsuarez, Jerryobject, Flyer22, Pemmy, Wrpscott, Agilemolecule, Beetstra public, VQuakr, Drtibbles, Vriend, Sellersb, WikHead, ElaineMeng, Valkerri, Yobot, Nicolazonta, Shadowboy813, Karnamohit, Citation bot, Gsmith8, Radiometer, Ludx, P99am, Anxdo, Mijuva, Catanist, FrescoBot, Citation bot 1, Sekmi, Vizbi, My very best wishes, Mittinatten, RjwilmsiBot, John of Reading, Rivanvx, Kajdhav, Williamseanohlinger, Vilietha, Alimay1985, Niko pd, Alpeshmalde, Cudauser, Helpful Pixie Bot, Northamerica1000, Kadhgar, Joseph Tucker Cambridge, Qgauss, Den2042, Mogism, Balder lai, Cantukarol, Lastawka and Anonymous: 68

- **Molecular design software** *Source:* https://en.wikipedia.org/wiki/Molecular_design_software?oldid=667373067 *Contributors:* Michael Hardy ,Jonabbey, Ghutchis, Quiddity, Beetstra, Xebvor, Spud Gun, JamesBWatson, Urbanrenewal, RosaWeber, Drtibbles, Rainman v84~enwik i,Lostraven, TutterMouse, Yobot, Materialscientist, Gsmith8, Ludx, P99am, Mijuva, FrescoBot, Palanimoe, EVTarau, Empty Buffer, Williams eanohlinger,Vzoete, Waghyogesh, Logasim, Cantukarol, Lifescienceinfo, ProfMichaelSternberg, Scienomics and Anonymous: 24

- **Molecular modelling** *Source:* https://en.wikipedia.org/wiki/Molecular_modelling?oldid=667296481 *Contributors:* Michael Hardy, Lexor, BenFrantzDale, Bensaccount, Unconcerned, Jason Quinn, Utcursch, LiDaobing, Karol Langner, Icairns, Rich Farmbrough, Cacycle, Enric Naval, TheParanoidOne, Hu, Richard Arthur Norton (1958-), Joerg Kurt Wegner, Bbullot~enwiki, YurikBot, Splette, Gaius Cornelius, Lijealso, Gadget850, DeadEyeArrow, Ms2ger, Itub, SmackBot, Weiguxp, Ohnoitsjamie, Kazkaskazkasako, Pieter Kuiper, Bduke, Para, Petermr, DMacks, Akpakp, Scottie 000, Dicklyon, Billgunn, Hu12, Bspahh, CmdrObot, Amalas, Van helsing, Gogo Dodo, Anonymi, Jamitzky, Headbomb, Trevyn, Second Quantization, Xebvor, AntiVandalBot, BokicaK, Ohlinger, Bikadi, JaGa, Zmm~enwiki, Uvainio, Nono64, Transisto, Maurice Carbonaro, Hodja Nasreddin, Nscerqueira, Lantonov, ACBest, Lordvolton, Manulinho72, ProteusCoop, MCTales, RosaWeber, SieBot, YAYsocialism, Nubiatech, Keilana, Shura58, Chenmengen, SimonTrew, Wikimcmd, Agilemolecule, Ideal gas equation, Drmolecule, Niceguyedc, Drtibbles, Rainman v84~enwiki, Puppy8800, Huyie, Graik, Excirial, Vriend, ElaineMeng, Addbot, Anotherdendron, Lightbot, Yobot, Nicolazonta, 豬豬, Materialscientist, Edguy99, Citation bot, Apothecia, Bio-ITWorld, P99am, Valeryns, Some standardized rigour, Catanist, FrescoBot, Jhfortier, Citation bot 1, Remyrem1, Cookiemonsier, Tbalius, RjwilmsiBot, Bento00, Wojcz, Math-ghamhainn, MithrandirAgain, Williamseanohlinger, Waghyogesh, Helpful Pixie Bot, Logasim, ThermoDynamicsUPB, Che kid, Compsim, Datta research, Sbalfour, Epicgenius, Cnmolak, YiFeiBot, Lizia7, Internucleotide, Anand guddu and Anonymous: 94

- **Molecular orbital** *Source:* https://en.wikipedia.org/wiki/Molecular_orbital?oldid=674735683 *Contributors:* Marj Tiefert, Josh Grosse, Michael Hardy, Kku, EdH, BRG, Smack, Malbi, Andrevan, Hao2lian, Ozuma~enwiki, Taxman, Donarreiskoffer, Gentgeen, Giftlite, Bensaccount, Chowbok, Karol Langner, Rdsmith4, Themusicking, Cfailde, Vsmith, Spoon!, Rcsheets, Nsaa, Benjah-bmm27, Wtmitchell, HenkvD, Dirac1933, Linas, Duncan.france, Ian**, SDC, Pfalstad, Graham87, V8rik, Nanite, Saperaud~enwiki, JHMM13, Fred Bradstadt, Terrx, Alphachimp, Chobot, YurikBot, Wavelength, SpuriousQ, GeeJo, Tetracube, Cobblet, Sbyrnes321, Itub, SmackBot, Stepa, Jrockley, Ohnoitsjamie, Isaac Dupree, Hugo-cs, Bduke, Colonies Chris, Drphilharmonic, DMacks, SteveLower, Sadi Carnot, Torritorri, Mgiganteus1, Twas Now, RSido, CmdrObot, Christian75, Thijs!bot, Nonagonal Spider, Headbomb, AntiVandalBot, Adams13, JAnDbot, MER-C, Jbom1, Burga, Baccyak4H, Dirac66, User A1, Bbi5291, CommonsDelinker, Pbroks13, Mikek999, MITBeaverRocks, Shoessss, Wyvern642, A4bot, Runewiki777, SieBot, Cwkmail, Smenge32, BartekChom, AghastAmok, ProkopHapala, Laburke, ClueBot, The Thing That Should Not Be, Puppy8800, Djr32, Nepenthes, WikHead, PL290, Addbot, AkhtaBot, Ronhjones, Orange Carrot, EconoPhysicist, Tide rolls, Javanbakht, Luckas-bot, Yobot, AnomieBOT, ^musaz, Reality006, RibotBOT, FrescoBot, Pinethicket, Kyr3915, RedBot, Bhxho, Double sharp, Lotje, Cosmo$angeet, KatelynJohann, EmausBot, Mnkyman, Superdelocalizable, ChuispastonBot, Tekstovi, Whoop whoop pull up, ClueBot NG, Wikiphysicsgr, UTChem40 ,MerllwBot, Bibcode Bot, Gluonman, BG19bot, BattyBot, Saehry, Lugia2453, Monkbot, Y-S.Ko, Arose2008, Scipsycho and Anonymous: 93

- **World Wide Molecular Matrix** *Source:* https://en.wikipedia.org/wiki/World_Wide_Molecular_Matrix?oldid=585880147 *Contributors:* Auric, S.K., Wavelength, Bluebot, Petermr, JoeBot, Kean1234, CmdrObot, Tomos ANTIGUA Tomos~enwiki, Michaeldsuarez, Bsherr, JL-Bot, 1ForTheMoney, Addbot, Download, Lightbot, Yobot, J04n, Paine Ellsworth, Lawsonstu and Anonymous: 2

25.3.2   Images

- **File:080998_Universe_Content_240.jpg** *Source:* https://upload.wikimedia.org/wikipedia/commons/a/a5/080998_Universe_Content_240.jpg *License:* Public domain *Contributors:* http://map.gsfc.nasa.gov/media/080998/index.html *Original artist:* Credit: NASA / WMAP Science Team

- **File:3GF1_Insulin-Like_Growth_Factor_Nmr_10_01.png** *Source:* https://upload.wikimedia.org/wikipedia/commons/7/70/3GF1_Insulin-Like_Growth_Factor_Nmr_10_01.png *License:* CC-BY-SA-3.0 *Contributors:* Self created from PDB entry with Cn3D Data Source: http://www.ncbi.nlm.nih.gov/Structure/ *Original artist:* Nevit Dilmen

- **File:3LRI_SolutionStructureAndBackboneDynamicsOfHumanLong_arg3_insulin-Like_Growth_Factor_1_02.png** *Source:* https://upload.wikimedia.org/wikipedia/commons/4/46/3LRI_SolutionStructureAndBackboneDynamicsOfHumanLong_arg3_insulin-Like_Growth_Factor_1_02.png *License:* CC-BY-SA-3.0 *Contributors:* Self created from PDB entry with Cn3D Data Source: http://www.ncbi.nlm.nih.gov/Structure/ *Original artist:* Nevit Dilmen

- **File:8tim_TIM_barrel.png** *Source:* https://upload.wikimedia.org/wikipedia/commons/3/3f/8tim_TIM_barrel.png *License:* CC-BY-SA-3.0 *Contributors:* Originally from en.wikipedia; description page is/was here. *Original artist:* Original uploader was WillowW at en.wikipedia

- **File:ATP-xtal-3D-sticks.png** *Source:* https://upload.wikimedia.org/wikipedia/commons/d/d4/ATP-xtal-3D-sticks.png *License:* Public domain *Contributors:* Own work *Original artist:* Ben Mills

- **File:AX4E0-3D-balls.png** *Source:* https://upload.wikimedia.org/wikipedia/commons/f/f4/AX4E0-3D-balls.png *License:* Public domain *Contributors:* Own work *Original artist:* Benjah-bmm27

- **File:AX6E0-3D-balls.png** *Source:* https://upload.wikimedia.org/wikipedia/commons/5/52/AX6E0-3D-balls.png *License:* Public domain *Contributors:* Own work *Original artist:* Benjah-bmm27

- **File:AX9E0-3D-balls.png** *Source:* https://upload.wikimedia.org/wikipedia/commons/4/4d/AX9E0-3D-balls.png *License:* Public domain *Contributors:* Own work *Original artist:* Ben Mills

- **File:Acetaldehyde-3D-vdW.png** *Source:* https://upload.wikimedia.org/wikipedia/commons/8/8e/Acetaldehyde-3D-vdW.png *License:* Public domain *Contributors:* ? *Original artist:* ?

- **File:Acetic-acid-3D-vdW.png** *Source:* https://upload.wikimedia.org/wikipedia/commons/d/db/Acetic-acid-3D-vdW.png *License:* Public domain *Contributors:* ? *Original artist:* ?

- **File:Affinity-table.jpg** *Source:* https://upload.wikimedia.org/wikipedia/commons/e/ee/Affinity-table.jpg *License:* Public domain *Contributors:* Table des differens rapports observes en chemie entre differentes substances; Memoires de l'Academie Royale des Sciences, pp. 202-212 *Original artist:* E.R. Geoffroy

- **File:Alchemy_air_symbol.svg** *Source:* https://upload.wikimedia.org/wikipedia/commons/b/b0/Alchemy_air_symbol.svg *License:* Public domain *Contributors:* Own work *Original artist:* Bryan Derksen

- **File:Alchemy_earth_symbol.svg** *Source:* https://upload.wikimedia.org/wikipedia/commons/1/16/Alchemy_earth_symbol.svg *License:* Public domain *Contributors:* Own work *Original artist:* Bryan Derksen

- **File:Alchemy_fire_symbol.svg** *Source:* https://upload.wikimedia.org/wikipedia/commons/4/4c/Alchemy_fire_symbol.svg *License:* Public domain *Contributors:* Own work *Original artist:* Bryan Derksen

- **File:Alchemy_water_symbol.svg** *Source:* https://upload.wikimedia.org/wikipedia/commons/0/0b/Alchemy_water_symbol.svg *License:* Public domain *Contributors:* Own work *Original artist:* Bryan Derksen

- **File:Aluminium_sulfate.jpg** *Source:* https://upload.wikimedia.org/wikipedia/commons/d/d3/Aluminium_sulfate.jpg *License:* Public domain *Contributors:* ? *Original artist:* ?

- **File:Ambox_important.svg** *Source:* https://upload.wikimedia.org/wikipedia/commons/b/b4/Ambox_important.svg *License:* Public domain *Contributors:* Own work, based off of Image:Ambox scales.svg *Original artist:* Dsmurat (talk · contribs)

- **File:Anthrax_toxin_protein_key_motif.svg** *Source:* https://upload.wikimedia.org/wikipedia/commons/4/4f/Anthrax_toxin_protein_key_motif.svg *License:* CC-BY-SA-3.0 *Contributors:* File:Anthrax toxin protein key motif.jpg *Original artist:*

- Original work: w:User:Natelewis

- **File:Asterisks_one.svg** *Source:* https://upload.wikimedia.org/wikipedia/commons/4/49/Asterisks_one.svg *License:* CC BY-SA 3.0 *Contributors:* Own work *Original artist:* DePiep

- **File:Asterisks_two.svg** *Source:* https://upload.wikimedia.org/wikipedia/commons/3/3f/Asterisks_two.svg *License:* CC BY-SA 3.0 *Contributors:* Own work *Original artist:* DePiep

- **File:Atisane3.png** *Source:* https://upload.wikimedia.org/wikipedia/commons/a/a4/Atisane3.png *License:* CC-BY-SA-3.0 *Contributors:* en:Image:Atisane3.png, en:Image:Atisane.png (molecule on the left) *Original artist:* en:User:Unconcerned, en:User:Ddoherty

- **File:Barium_unter_Argon_Schutzgas_Atmosphäre.jpg** *Source:* https://upload.wikimedia.org/wikipedia/commons/1/16/Barium_unter_Argon_Schutzgas_Atmosph%C3%A4re.jpg *License:* Public domain *Contributors:* yes *Original artist:* Matthias Zepper

- **File:Bent-3D-balls.png** *Source:* https://upload.wikimedia.org/wikipedia/commons/4/47/Bent-3D-balls.png *License:* Public domain *Contributors:* ? *Original artist:* ?

- **File:Benz1.png** *Source:* https://upload.wikimedia.org/wikipedia/commons/9/9c/Benz1.png *License:* CC-BY-SA-3.0 *Contributors:* ? *Original artist:* ?

- **File:Beta-meander1.png** *Source:* https://upload.wikimedia.org/wikipedia/commons/b/b8/Beta-meander1.png *License:* Public domain *Contributors:* Transfered from en.wikipedia *Original artist:* Original uploader was Xenonblast at en.wikipedia

- **File:Boltzmanns-molecule.jpg** *Source:* https://upload.wikimedia.org/wikipedia/commons/2/20/Boltzmanns-molecule.jpg *License:* Public domain *Contributors:* Transfered from en.wikipedia Transfer was stated to be made by User:Blast. *Original artist:* Original uploader was Sadi Carnot at en.wikipedia

- **File:Bromine_vial_in_acrylic_cube.jpg** *Source:* https://upload.wikimedia.org/wikipedia/commons/3/35/Bromine_vial_in_acrylic_cube.jpg *License:* CC BY-SA 3.0 de *Contributors:* Own work *Original artist:* Alchemist-hp (pse-mendelejew.de)

- **File:Buckminsterfullerene-perspective-3D-balls.png** *Source:* https://upload.wikimedia.org/wikipedia/commons/0/0f/Buckminsterfulleralls.png *License:* Public domain *Contributors:* Own work *Original artist:* Benjah-bmm27

- **File:Butan_Lewis.svg** *Source:* https://upload.wikimedia.org/wikipedia/commons/c/cb/Butan_Lewis.svg *License:* Public domain *Contributors:* Own work *Original artist:* NEUROtiker ⇌

- **File:Carbon-monoxide-3D-vdW.png** *Source:* https://upload.wikimedia.org/wikipedia/commons/a/a7/Carbon-monoxide-3D-vdW.png *License:* Public domain *Contributors:* ? *Original artist:* ?

- **File:Carboxylic_acid_dimers.png** *Source:* https://upload.wikimedia.org/wikipedia/commons/c/c9/Carboxylic_acid_dimers.png *License:* Public domain *Contributors:* ? *Original artist:* ?

- **File:Ch4_hybridization.svg** *Source:* https://upload.wikimedia.org/wikipedia/commons/b/bb/Ch4_hybridization.svg *License:* Public domain *Contributors:* Transferred from en.wikipedia
Original artist: Original uploader was K. Aainsqatsi at en.wikipedia

- **File:Chem_template.svg***Source:*https://upload.wikimedia.org/wikipedia/commons/a/ac/Chem_template.svg*License:*Public domain*Con-tribut ors:*own work inspired by <a href='//commons.wikimedia.org/wiki/File:Chem_template.png' class='image'><img alt='Chem template.png'src='ht tps://upload.wikimedia.org/wikipedia/commons/6/64/Chem_template.png' width='48' height='48' class='thumbborder' data-file-width='48'data-file-height='48' /></a><a href='//commons.wikimedia.org/wiki/File:Chem_template.svg' class='image'><img alt='Chem template.svg'src='htt ps://upload.wikimedia.org/wikipedia/commons/thumb/a/ac/Chem_template.svg/46px-Chem_template.svg.png' width='46' height='46'class='th umbborder' srcset='https://upload.wikimedia.org/wikipedia/commons/thumb/a/ac/Chem_template.svg/69px-Chem_template.svg.png1.5x, https ://upload.wikimedia.org/wikipedia/commons/thumb/a/ac/Chem_template.svg/92px-Chem_template.svg.png 2x' data-file-width='400'data-file-h eight='400' /></a>*Original artist:*Amada44

- **File:Cobalamin.png** *Source:* https://upload.wikimedia.org/wikipedia/commons/e/ec/Cobalamin.png *License:* Public domain *Contributors:* Transferred from en.wikipedia
Original artist: Ymwang42 (talk). Original uploader was Ymwang42 at en.wikipedia

- **File:Combustion_reaction_of_methane.jpg***Source:*https://upload.wikimedia.org/wikipedia/commons/7/7c/Combustion_reaction_of_met hane.jpg *License:* CC BY-SA 3.0 *Contributors:*

- Methane-3D-space-filling.svg *Original artist:*

- Jynto

- **File:Commons-logo.svg** *Source:* https://upload.wikimedia.org/wikipedia/en/4/4a/Commons-logo.svg *License:* ? *Contributors:* ? *Original artist:* ?

- **File:Copper.jpg***Source:*https://upload.wikimedia.org/wikipedia/commons/f/fb/Copper.jpg*License:*CC BY 3.0*Contributors:*http://images-of-el ements.com/copper.php *Original artist:* Jurii

- **File:Coupers-molecule.jpg** *Source:* https://upload.wikimedia.org/wikipedia/commons/d/d6/Coupers-molecule.jpg *License:* Public domain *Contributors:* Transferred from en.wikipedia to Commons by Hystrix using CommonsHelper. *Original artist:* The original uploader was Sadi Carnot at English Wikipedia

- **File:Crystal_Clear_device_cdrom_unmount.png** *Source:* https://upload.wikimedia.org/wikipedia/commons/1/10/Crystal_Clear_device_ cdrom_unmount.png *License:* LGPL *Contributors:* All Crystal Clear icons were posted by the author as LGPL on kde-look; *Original artist:* Everaldo Coelho and YellowIcon;

- **File:Cubical_atom_2.png** *Source:* https://upload.wikimedia.org/wikipedia/commons/1/14/Cubical_atom_2.png *License:* Public domain *Contributors:* Taken from JACS article by G. N. Lewis published in 1916 (see [1]) *Original artist:* G.N. Lewis

- **File:Cyanooctatetrayne-3D-vdW.png** *Source:* https://upload.wikimedia.org/wikipedia/commons/f/f7/Cyanooctatetrayne-3D-vdW.png *License:* Public domain *Contributors:* Derived from File:Acetylene-3D-vdW.png and File:Hydrogen-cyanide-3D-vdW.png. *Original artist:* Ben Mills and Jynto

- **File:D-Fructose_cyclic.png** *Source:* https://upload.wikimedia.org/wikipedia/commons/6/61/D-Fructose_cyclic.png *License:* CC BY 3.0 *Contributors:* Own work *Original artist:* Physchim62

- **File:D-tartaric_acid.png** *Source:* https://upload.wikimedia.org/wikipedia/commons/b/b8/D-tartaric_acid.png *License:* Public domain *Contributors:* Transferred from en.wikipedia to Commons. *Original artist:* The original uploader was Mykhal at English Wikipedia

- **File:DGlucose_Fischer.svg** *Source:* https://upload.wikimedia.org/wikipedia/commons/1/14/DGlucose_Fischer.svg *License:* CC BY-SA 3.0 *Contributors:* Own work *Original artist:* Christopher King

- **File:DIMendeleevCab.jpg** *Source:* https://upload.wikimedia.org/wikipedia/commons/c/c8/DIMendeleevCab.jpg *License:* Public domain *Contributors:* Transferred from ru.wikipedia
Original artist: −. Original uploader was Serge Lachinov at ru.wikipedia

- **File:DIN_4844-2_Warnung_vor_Laserstrahl_D-W010.svg** *Source:* https://upload.wikimedia.org/wikipedia/commons/1/16/DIN_4844-2_ Warnung_vor_Laserstrahl_D-W010.svg *License:* Public domain *Contributors:* Own work *Original artist:* Torsten Henning

- **File:Dalton_John_desk.jpg** *Source:* https://upload.wikimedia.org/wikipedia/commons/3/3f/Dalton_John_desk.jpg *License:* Public domain *Contributors:* Frontispiece of *John Dalton and the Rise of Modern Chemistry* by Henry Roscoe *Original artist:* Henry Roscoe (author), William Henry Worthington (engraver), and Joseph Allen (painter)

- **File:Daltons_particles.jpg** *Source:* https://upload.wikimedia.org/wikipedia/commons/2/23/Daltons_particles.jpg *License:* Public domain *Contributors:* File Daltons-particles.jpg found at History of molecular theory *Original artist:* John Dalton; published in: *New System of Chemical Philosophy*

- **File:Dendrimer_ChemEurJ_2002_3858.jpg** *Source:* https://upload.wikimedia.org/wikipedia/commons/4/4a/Dendrimer_ChemEurJ_2002_ 3858.jpg *License:* CC-BY-SA-3.0 *Contributors:* Transferred from en.wikipedia to Commons. *Original artist:* M stone at English Wikipedia

25.3.3   Content license